NEW SURVEY OF CLARE ISLAND

Volume 4: The Abbey

Edited by
Conleth Manning
Paul Gosling
and John Waddell

Royal Irish Academy

First published in 2005
by the Royal Irish Academy,
19 Dawson St, Dublin 2.
http://www.ria.ie
Copyright © Royal Irish Academy 2005

Figures and maps based on Ordnance Survey Ireland by permission of Ordnance Survey Ireland (Permit No. 7892) © Ordnance Survey Ireland and Government of Ireland

British Cataloguing-in-Publication Data.
A catalogue record for this book is available from the British Library.

ISBN 1-904890-05-9

Design (based on Volume 1) by Vermillion Design
Typesetting by Phototype-Set Ltd, Dublin
Printed in Spain by Estudios Gráficos Zure, Bilbao

This publication has received support from

the Department of the Environment,
Heritage and Local Government

the Heritage Council
under the 2005 Publications Grant Scheme

the Royal Society of Antiquaries of Ireland
(Bevir Trust Archaeological Fund)

CONTENTS

NEW SURVEY OF CLARE ISLAND
Committee Members and Fieldworkers

New Survey of Clare Island Executive Committee 2004–2005

J. Breen
P. Coxon
G.J. Doyle (Secretary, 1991–2004)
D.J. Fegan (Science Secretary, 2004–)
J.R. Graham
M.D.R. Guiry

T. Kelly
R.P. Kernan (Chairman, 1997–2004)
C. Mac Cárthaigh
T.K. McCarthy
C. Manning (Deputy Secretary, 1996–2004; Secretary 2004–)

A.A. Myers
M.E.F. Ryan (President, 2002–)
M.W. Steer (Managing Editor, 1997–; Chairman, 2004–)

Former committee members: A. Clarke (President, 1990–3), D. Cabot (Chairman, 1989–97), J.S. Fairley, M. Herity (President, 1996–9), G.F. Imbusch, W.I. Montgomery, J.O. Scanlan (President, 1993–6), J.A. Slevin (Science Secretary, 2000–2004), T. D. Spearman (President, 1999–2002), P.D. Sweetman, J. Waddell, K. Whelan.

Fieldworkers and other contributors

Botany

J. Bryant	H. Fuller	R. McHugh	F. Rindi	P. Sims
E.J. Cox	M.D.R. Guiry	C.A. Maggs	T. Ryle	D. Synnott
G.J. Doyle	P.W. James	D.H.S. Richardson	M.R.D. Seaward	R.S. ten Cate
H. Fox	D.M. John			

Zoology

P. Allen	R.E. Cussen	R. Grabda	A. Macfadyen	R. O'Riordan
R. Anderson	J. Dawes	L. Harrington	D. McFerran	C. O'Toole
T. Bolger	E. de Eyto	M. Hartnett	M. McGarrigle	B. Picton
K. Bond	J. Delany	S. Hassett	F. McGowan	A.M. Power
J. Breen	J. Dick	A. Hopkirk	D. McGrath	J. Reynolds
D. Cabot	R. Edwards	M. Hynes	E. MacLoughlin	E. Rogan
R. Caroni	S. Fahy	B. Keegan	N. McLoughlin	C. Sheils
C. Carter	L. Ferron	B. Keely	F. Marnell	P. Smiddy
M. Cawley	P. Fitzmaurice	T. Kelly	W.I. Montgomery	D. Tangney
P. Chandler	J. Fives	M. Kelly-Quinn	D. Murray	P. Tattersfield
M.J. Costello	M. Flanagan	R. Kennedy	A.A. Myers	F. Tiernan
J. Costelloe	I. Forsyth	G. Konnecker	J. Nunn	C. Tobin
D. Cotton	G.N. Foster	E. Lalloway	D. Nowak	A. Trojanowska
K. Creed	T. Gallagher	A. Lawless	I. O'Connor	P.M. Walsh
M. Cronin	J. Galvin	J. Lusby	G. O'Donnell	M. Zalewski
T. Cross	F. Geraghty	K. McAney	S. O'Driscoll	G. Walsh-Kemmis
P. Cullen	P. Giller	T.K. McCarthy	D.B. O'Leary	N.J. Whitehouse
C. Cunnane	J. Good	S. McCormack		

Geology

A. Brock	G.A. Doyle	K. Higgs	P. Morris	C.J. Stillman
C. Brown	D. Evans	P. Lynch	G. Petrie	B.G.J. Upton
J.F. Collins	J.R. Graham	M. McCarthy	W.E.A. Phillips	W. Vullings
R. Corcoran	J. Harkin	B. Methley	G.W. Smillie	M. Williams
P. Coxon	D. Harper			

Archaeology

K. Barton	J. Fenwick	B. Kulessa	K. Morton	B. Ronayne
A. Buckley	F. Gillespie	H. Lavelle	M. Ó Comáin	L. Slater
I. Cantwell	P. Gosling	C. Manning	C. Oldenbourg	R. Stalley
M. Casey	M. Katkov	D. Monaghan	E. Ramos Julian	P.D. Sweetman
M. Comber	P. King			

History, Place-Names and Folklife

T. Collins	S. Ó Catháin	N. Ó Muraíle
C. Mac Cárthaigh	A. Ó Maolfabhail	K. Whelan

ACKNOWLEDGEMENTS

The production of the fourth volume of the New Survey of Clare Island would not have been possible without the generous sponsorship and support of the National Monuments Section, Department of the Environment, Heritage and Local Government, the Heritage Council and the Bevir Trust. Much of the new information in this volume on the Abbey and in particular on its paintings is the result of a major conservation project on the paintings in the Abbey undertaken by the National Monuments Service. We would like to thank the Senior Architects in charge of this project, Willie Cumming and Paul McMahon; the clerk-of-works, John Corcoran; the contracted conservators, who carried out the work, Christoph Oldenbourg and Madeleine Katkov; all who worked on that project, especially Karena Morton; and all of the contributors to this volume.

The New Survey of Clare Island would not have become a reality without the continued support and cooperation of all the people and institutions who have given generously of their time and expertise during the past twelve years. Particular mention should be made of the current and former members of the Executive Committee for their ongoing commitment to the survey. Special thanks must be given this year to the former Chairman and Secretary, Professor R.P. Kernan and Professor G. Doyle, whose four-year terms came to an end in 2004. Thanks are also due to fieldworkers, students, academics and professionals from all over Ireland and the world for their generous contributions in terms of their time and expertise and for braving the crossing to Clare Island.

We would like to take this opportunity to express our appreciation of the support provided by all of our sponsors: Enterprise Energy Ireland; the Marine Institute; Mayo County Council; the Heritage Council; the Discovery Programme Ltd; the Department of the Environment, Heritage and Local Government (National Parks and Wildlife Section and Environment Awareness Awards, 1992–7); the Royal Irish Academy Bicentennial Research Trust; CRH plc; the Marine Institute; the British Council; eircom; the Central Fisheries Board; ESB—Northwest Region; and Roankabin Ltd.

The many individuals who have made donations to help fund the work of the New Survey deserve thanks (sadly, some have since died[†]): [†]Professor Brendan Keegan, J. Bradley, J.R. Creighton, M. Davies, P. and S. Ewen, J.S. Fairley, C.J. Haughey, [†]J. Heslop-Harrison, G.L. Huxley, R.C. Lewis Crosby, G. McCall, N. McCann, [†]G.F. Mitchell, Mrs. P. Moran, [†]Lord Moyne, M. Mulcahy, J. and S. Mulloy, M. O'Donnell, T. O'Malley Seidler, C. O'Rourke, H. O'Shea, C.S. Ronayne, M.J.P. Scannell, [†]E. Twomey, F. Ward-O'Malley, and G.T. Wrixon.

Many institutions supported the research, fieldwork and preparation of results for publication: the National University of Ireland, Galway; Queen's University Belfast; Trinity College, Dublin; University College Cork (NUI, Cork); University College Dublin (NUI, Dublin); the Department of the Environment, Heritage and Local Government; the Galway/Mayo Institute of Technology; the National Botanic Gardens; the National Library of Ireland; the National Museum of Ireland; the Natural History Museum, London; the Ulster Museum, Belfast; the Metereological Service, Dublin; the Office of Public Works; the Ordnance Survey of Ireland; Mayo County Council and the Regional Water Laboratory, Castlebar.

The Academy's staff and members are to be sincerely thanked for their dedication and expertise.

Finally, our thanks are due to the people of Clare Island, Co. Mayo, who made our time on the island memorable. Special thanks go to Peter and Anna Gill, Ciara Cullen, Oliver and Mary O'Malley, Bernie Winters, Charles O'Malley and Chris O'Grady.

Martin Steer
Managing Editor,
New Survey of Clare Island series

NOTES ON CONTRIBUTORS

Ann Buckley is a Government of Ireland Senior Research Fellow in the Music Department, National University of Ireland, Maynooth, and a Research Associate of the Centre for Medieval and Renaissance Studies, Trinity College, Dublin. Author of the chapter 'Music in Ireland *c*. 1500' in *A New History of Ireland*, Vol. 1 (Oxford, forthcoming), she has published numerous articles and lectured widely on the topic of music in medieval Ireland.

Ian Cantwell is a researcher and landscape historian with an interest in the systematic recording of gravestone transcriptions. His most recent publication is an interactive CD-ROM of gravestone inscriptions in the west of Ireland entitled *Memorials of the Dead: Counties Galway & Mayo (Western Seaboard)*.

Fergus Gillespie is Deputy Chief Herald at the Office of the Chief Herald. He is Vice-President of the Society of Genealogists, England, an Officer of the *Bureau Permanent* of the International Congress of Genealogical and Heraldic Sciences, Fellow of the Argentinian Genealogical Society, an associate member of the *Académie Internationale d'Heraldique* and a Fellow of the *Real Academia Matritense de Heráldica y Genealogía*. He has written and lectured widely on early Irish language, literature and culture, as well as on Irish heraldry and early Gaelic and Norman genealogies.

Paul Gosling is an archaeologist and lecturer in the Department of Humanities, Galway–Mayo Institute of Technology, where he co-ordinates the diploma/degree programmes in Heritage Studies. From 1984 to 1993, he was director of the Office of Public Works/ University College Galway archaeological survey of County Galway. He also directed the archaeological element of the New Survey of Clare Island. His publications include the archaeological inventories of County Galway and a number of books and papers on Irish towns.

Conleth Manning has worked for twenty-five years as an archaeologist with the National Monuments Service, presently part of the Department of the Environment, Heritage and Local Government. He has published many papers relating to Irish archaeology, is particularly interested in the medieval period and is currently President of the Royal Society of Antiquaries of Ireland.

Karena Morton is a conservator of archaeological finds and wall paintings. She is a graduate of University College Dublin and the University of Wales, College of Cardiff, and is currently working for a PhD on Irish medieval wall paintings at the School of Architecture, University College Dublin. She has directed or co-directed a number of wall painting conservation programmes, including those at Ardamullivan Castle, Co. Galway, Barryscourt Castle, Co. Cork and Ballyportry Castle, Co. Clare, and she has written a number of articles on medieval wall paintings in Ireland.

Micheál Ó Comáin is a Herald of Arms at the Office of the Chief Herald of Ireland and a member of the *Bureau Permanent* of the International Congress of Genealogical and Heraldic Sciences. He had written and lectured extensively on Irish Armory.

Christoph Oldenbourg is a conservator of architectural decoration, polychromy and gilding, specialising in wall paintings of Roman, medieval and more recent objects. He has worked on and directed wall painting conservation projects across Europe and in America. In Ireland he has co-directed the wall painting conservation projects at Abbeyknockmoy and on Clare Island and has worked on a number of other projects. He has published a number of articles on wall painting conservation.

Roger Stalley is Professor of the History of Art at Trinity College, Dublin. He has written and lectured widely on Irish medieval art and architecture and is author of several books, including *The Cistercian Monasteries of Ireland* (1987) and *Early Medieval Architecture* (1999).

John Waddell is Professor of Archaeology, National University of Ireland, Galway. His publications include studies of Irish early Bronze Age pottery and burials and a major synthesis of the archaeology of prehistoric Ireland. Fieldwork has included work on the Aran Islands and in the royal site of Rathcroghan, Co. Roscommon. He is a member of the Royal Irish Academy.

A view of the painted vaulted ceiling and east wall of the chancel of the Abbey, Clare Island, from the west (Con Brogan, DOEHLG).

INTRODUCTION

Conleth Manning

The Abbey on Clare Island is a unique survival among Irish medieval churches because of its extraordinary wall paintings. No other medieval church in Ireland gives nearly as good an impression of what a medieval painted interior was like. In the next best example, Cormac's Chapel at Cashel, Co. Tipperary, the paintings are more fragmentary and today do not dominate the chancel in the same way.

This Cistercian establishment on Clare Island would probably never have qualified as an abbey because it would not have been able to support the necessary complement of twelve monks. Whatever the intentions of its founders, all of the scant historical information indicates that it was a cell or subsidiary house attached to the abbey of Knockmoy, Co. Galway (Manning, this volume, fig. 2, p. 8). Though not a full abbey, it was nevertheless the most westerly presence of the Cistercians in Europe and the only Irish Cistercian house situated on an island.

The history and archaeology section in the original Clare Island Survey was written by Thomas Johnson Westropp (1860–1922) (Pl. I). He qualified from Trinity College, Dublin, as a civil engineer and worked for some three years in this profession but, as he had sufficient private means, he was thereafter (from 1888) able to devote himself full time and unpaid to his antiquarian interests (Ashe Fitzgerald 2000, 15–17). He was elected a member of the Royal Irish Academy in 1894, served as president of the Royal Society of Antiquaries of Ireland from 1916 to 1919 and published over one hundred articles, many of which were about field monuments, churches and castles in counties Clare and Limerick. He also made a special study of ring forts, stone forts and, in particular, promontory forts. He was a competent draftsman and artist and illustrated his articles with his own drawings. At the time of the original Clare Island Survey he was the ideal person to record the archaeology of the island and the Abbey and its wall paintings in particular.

As with many of the fieldworkers in the original survey, Westropp ranged wider than the confines of the island itself to describe the archaeology of the neighbouring islands. He devoted more text and illustrations to the Abbey on Clare Island (nine pages of text and most of five full-page plates) than to any other site or monument type. His was the first detailed description and plan of the building and the first, and to date the most complete, published record of the mural paintings. He recorded the paintings with great thoroughness because he believed they would not last much longer (Westropp 1911, 33). A platform of barrels and planks was set up so that he could trace the paintings directly off the vaulted ceiling (Westropp 1911, 29). These original tracings survive in the collection of the Royal Society of Antiquaries of Ireland. From these tracings he made watercolour illustrations at a reduced scale in his notebooks, which are now preserved in the Royal Irish Academy. His notes, tracings and original illustrations combined with his published text form a unique record of the paintings as they were in the early twentieth century. This archive is particularly important because of the loss or deterioration of some of the images in the intervening years.

Just as Westropp's account of the Abbey is the longest entry on any site in his published text, the detailed account of the Abbey and its paintings presented here was considered of sufficient importance and interest to merit a separate volume in the New Survey of Clare Island series. This volume represents a major advance on Westropp's important pioneering work, which was undertaken under difficult circumstances with little or no technical back-up. The great advances in our knowledge of the wall paintings and, indeed, the tremendous improvement in their condition is due to the fact that the National Monuments Service had been engaged in a major conservation project on the paintings at the same time as the survey was in progress. As well as

Pl. I Thomas Johnson Westropp (1860–1922), author of the history and archaeology section of the original Clare Island Survey (Royal Society of Antiquaries of Ireland).

cleaning and securing the paintings and the plaster, this work has uncovered a number of images never seen before in modern times. The experts involved in the conservation describe here the pioneering scientific methods they employed to rescue this unique and extraordinary medieval work of art from further imminent deterioration. It is a boon to the survey to be able to present, with the cooperation of the National Monuments Service, the first detailed account of this important work.

The following contributions on the Abbey cover the building and its history, the paintings, their conservation and iconography, the musical instruments shown in the paintings, the armorial plaque, and the associated monuments and the graveyard inscriptions and offer a general discussion of the Abbey and its paintings. It should be pointed out that the building phases and the painting phases are separately numbered simply because there is one major building phase

that predates any surviving paintwork. To prevent confusion the painting phases are usually referred to as such and the numbers are written out in full, as Painting Phase One and Painting Phase Two. In the case of the building phases an arabic numeral is used, i.e. Phase 1, Phase 2, etc. Building Phase 2, the building of the chancel, is contemporary with Painting Phase One and it is thought that Building Phase 3, when the domestic wing was added, is contemporary with Painting Phase Two.

It has been difficult to fully standardise the spelling of the Irish personal names among the various papers in this volume. While it is a common practice in Irish historical writing to give Gaelic personal names in their roughly contemporary Irish-language form, some English forms are so frequently used in the art-historical literature, such as Cooey na Gall O'Cahan and Art Mac Murrough, that they have been retained here in these forms. The modern English form O'Malley is used here interchangeably with its Irish form, Ó Máille. As most of the members of the O'Conor family mentioned here are members or descendants of the family of the kings of Connacht, O'Conor, the form still used by this family, is used here rather than the more common modern form, O'Connor. However, Cathal Crobderg Ó Conchobair is written as in *A new history of Ireland*. Names in quoted text are, of course, left as found. Photographs are by the authors of the relevant papers unless otherwise stated.

The photographs of individual painted images and of Westropp's notebooks in the catalogue of the wall paintings (this volume, pp 61–95) are by Christoph Oldenbourg. The letters 'RIA' after the latter indicate that the Royal Irish Academy holds the copyright of these images.

The objective survey detail and illustrations in the following accounts will stand the test of time and, like Westropp's account and archive, will become part of the primary record. Full agreement among the many authors on every detail of interpretation and dating would be neither possible nor desirable, but it is hoped that the discursive elements in these accounts are a reasonable step towards the truth in as far as it will ever be determined. No doubt the opinions expressed here will and should be challenged and modified in due course, especially when neglected aspects of Irish medieval art and architecture are more fully surveyed and studied.

REFERENCES

Ashe Fitzgerald, M. 2000 *Thomas Johnson Westropp (1860–1922): an Irish antiquary*. Seandálaíocht: Monograph 1. Dublin. Department of Archaeology, University College, Dublin.

Westropp, T.J. 1911 Clare Island Survey: history and archaeology. *Proceedings of the Royal Irish Academy* **31** (1911–15), section 1, part 2, 1–78.

SECTION ONE
THE ABBEY, GRAVEYARD AND ASSOCIATED MONUMENTS

HISTORY, SURVEY AND ANALYSIS OF THE BUILDING

Conleth Manning

ABSTRACT

The medieval Cistercian cell on Clare Island, known as the Abbey, is described and discussed. A new survey and analysis of the building is presented along with an account of its history. Three main building phases are described, ranging from the thirteenth to the fifteenth centuries. It is argued that the chancel and its two phases of painting date from the fifteenth century.

Introduction

The building known as the Abbey is situated about midway along the southern side of Clare Island on a terrace on rising ground some 250m from the shore (Fig 1; Pls I–IV). That this was an early medieval church site is indicated by the existence of a tall cross-inscribed pillar here and an altar and holy well closeby, which is named Toberfelamurry on the Ordnance Survey map of 1838 (see Gosling, this volume, p. 33). The building consists of a church with nave and chancel and a small, narrow, ruined wing attached to the north side of the chancel. The chancel is a tall, tower-like two-storey structure with a vaulted first floor. The attached building was a lower two-storey domestic structure. The most notable feature of the entire building is the chancel, with its canopied wall tomb and remarkable mural paintings.

History

Nothing is recorded about this establishment apart from the fact that it was a dependent small house or cell attached to the Cistercian abbey of Knockmoy, Co. Galway. Sir James Ware, who had in his collection the now lost chartulary roll of the abbey of Knockmoy (O'Sullivan 1997, 71, 89), wrote that the church on Clare Island was annexed to Knockmoy sometime after the death

in 1224 of the founder of that abbey, Cathal Crobderg Ó Conchobair, king of Connacht (Ware 1626, 75–6; 1658, 252–3). Knockmoy was founded in 1190 as a daughter house of Boyle, which in turn was a daughter house of Mellifont (Fig. 2), the first Irish Cistercian abbey, founded in 1142 (Stalley 1987, 11–13, 240).

Malachy Hartry, an Irish Cistercian writing in 1640, expressed uncertainty as to whether Clare Island was an abbey or a priory (Murphy 1891, 199). Post-dissolution grants and surveys indicate that it belonged to Knockmoy (Westropp 1911, 15; Gwynn and Hadcock 1970, 124, 129). This is the only reliable historical information on the foundation—suggestions that it was ever a Carmelite cell (Westropp 1911, 14) or that it was 'built by the O'Mailles, in 1224, according to Ware' (Westropp 1911, 14) or by Grace O'Malley in the sixteenth century (Deane 1880, 75) or was founded (or rebuilt) in 1460 as Luckombe states (Stalley 1987, 243)[1] have to be rejected. The earliest reference to the paintings is from a work by Gaspar Jongelinus published in 1640, which is referred to by Alemand (1690, 190; 1722, 195). The relevant section (Jongelinus 1640, Liber viii, 30) is given here in the original Latin and in a translation kindly undertaken by the late Liam de Paor:

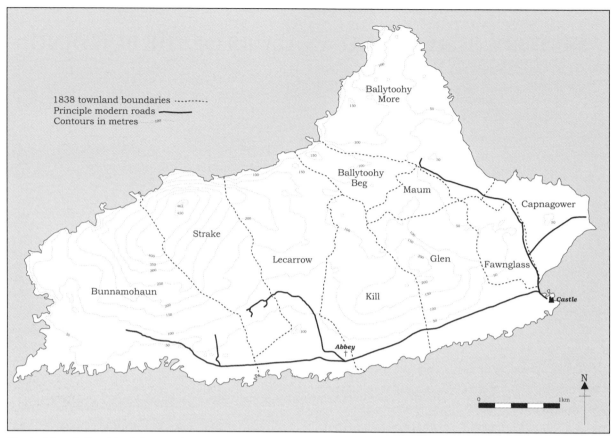

Fig. 1   Map of Clare Island showing location of the Abbey.

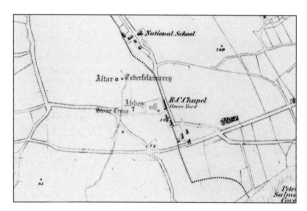

Pl. I   Ordnance survey first edition six–inch map of area around Abbey.

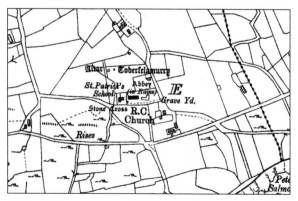

Pl. II   Ordnance survey six–inch map of area around Abbey (twentieth century).

Fig. 2   Map of Ireland showing location of medieval Cistercian houses.

Pl. III A view of the Abbey in May 1952, some months before the chancel was re-roofed (DOEHLG).

Pl. IV Aerial view of the area around the Abbey (2001) showing the well on the extreme right and the seashore in the background (Con Brogan, DOEHLG).

Porro adhuc supersunt in hac insula, qui recordantur se monachos huius loci vidisse & novisse quos Scoti Piratae inde ignominiose eiecerunt & quidquid ibidem erat, abstulerunt. Integrum tamen etiamnum remanet artificiose constructum & depictum.

[Moreover, Scottish pirates expelled ignominiously those who formerly remained on this island—who recount that they saw themselves as the monks of that place—and robbed everything that was there. However, the masonry structures, with their painted work, remain intact.]

Jongelinus acknowledged Ware, Hartry and Colgan in his introduction (1640, Liber viii, 20), but he may also have had up-to-date information on some of the Irish houses from Irishmen living on or visiting the Continent. While two works by Hartry have been edited, there were others that do not appear to have survived (Murphy 1891, 296–7). The account of the attack by the Scottish pirates probably came from some such source and is likely to refer to the 1580s or 1590s, when there were frequent raids on the west coast of Connacht by Scots from the Western Isles (Hayes-McCoy 1996, 142). It is recorded that Mac Neill of Barra raided O'Malley's Country on a number of occasions and that Mac Neill's lands were raided by Grace O'Malley. The account by Jongelinus was completely misunderstood in the past, and the attack by pirates was erroneously assigned to the thirteenth century (Gwynn and Hadcock 1970, 129; Stalley 1987, 243). Also from the seventeenth century there are references to the dissolved abbey and a half quarter of land belonging to it (Lecarrow or 'Lecarrownemanestragh' as it is called in the Book of Survey and Distribution) (O'Sullivan 1958, 39; Simington 1956, 111).

R. Downing, in a description of County Mayo dating from about 1684, claims that the 'abbey of St Bernard of Cliera' was built by 'Dermitius (Claudus)[2] O Maly' and that he and his wife, 'Maud Curvan f[...]O Connor' and most of his family up to the later seventeenth century were buried there (Ó Muraíle 1999, 104). This appears to be the same couple, 'Diarmuid Bacach O'Malley' and his wife 'Lady Maeve O'Conor', who were involved in the foundation of the Augustinian friary of Murrisk in 1456 (Ó Muraíle 1998, 256; Gwynn and Hadcock 1970, 300). As he was possibly dead and certainly no longer chief

by 1456, there is a possibility that he was the Diarmait Ó Máille who won the chieftainship in 1415, according to the Annals of Connacht (Freeman 1970, 426–7).

There is no record of when the church went fully out of use as a place of worship or when it became unroofed. I draw a distinction here because the celebration of mass could have continued in the vaulted chancel even if the nave was roofless, as happened at Ballintober Abbey, Co. Mayo. By the time of the first edition of the Ordnance Survey map of 1838, a long narrow chapel, aligned north–south along the roadway immediately east of the Abbey, was in existence (Pl. I). This is likely to have been in use since sometime in the eighteenth century, and it is possible that the bell cot on the east gable of the Abbey (Pl. III) was erected to serve this chapel. The present chapel to the west of the Abbey was built about 1862 (Westropp 1911, 33). The earliest published illustration of the Abbey, or rather part of it, is a print of the exterior of the east window (Pl. V) in Otway's Tour of Connaught (1839, 300). Though he described his visit to the Abbey he made no mention of the paintings and devoted

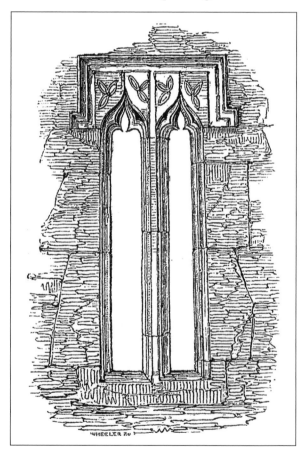

Pl. V   A print of the east window from Otway (1839).

most of his account to 'a skull with gold earrings' that the locals claimed was that of Grace O'Malley.

The Abbey was declared a National Monument (no. 97) in 1880 under the Church Temporalities Act, but the short account of the building in the report of the Commissioners of Public Works for that year does not even mention the paintings. Ownership of the monument and graveyard was transferred to the state by the Congested Districts Board in 1895. Around the time Westropp was recording the paintings the Office of Public Works removed the soil from the top of the vault and laid a cement floor over it in an attempt to keep it dry. This in fact did more damage than it prevented, and by the 1950s the cement floor had cracked and was leaking water onto the vault, accompanied now by harmful salts from the cement. About 1951 the renowned British conservator H.J. Plenderleith was brought to the island to advise on the paintings, and he recommended removing the cement floor and putting a roof on the chancel (Liam de Paor, pers. comm.). This was done in the winter of 1952 (Pls VI, VII, VIII, IX), and at the same time the interior of the building was cleared to the medieval floor level. The present campaign of conservation work on the building and on the paintings was initiated

in 1990, and as part of this work the entire two-storey chancel building was lime-rendered in 1997, in a successful attempt to dry out the structure. In a further attempt to protect the paintings, the nave was roofed in the summer of 2002 (Pl. X).

Description

There are three main phases in the building, the nave being the primary structure (Fig. 3). The two-storey chancel was added to the nave and finally the north wing was added to the chancel. The main elements of the building are described in the sequence in which they were built, and an analysis of the phases follows.

The nave (Fig. 3)

This plain rectangular structure is 11.15m long by 5.80m wide internally. The masonry of the walls includes many large unshaped stones, often rounded, and incorporates blocky corner stones where the thickness is often similar to the height. The main entrance is a doorway in the west wall that is slightly off-centre. Externally it has a segmental pointed head with an external hollow chamfer and triangular stop chamfers at the base. Internally the two sides of the doorway are very different. That to the south is at right angles to the

Pl. VI The Abbey from the west-north-west in 1952 (DOEHLG).

Pl. VII   The Abbey from the west in 1991.

Pl. VIII   The Abbey from the south-east (1970s).

Pl. IX   The Abbey from the south-east (1990s) (DOEHLG).

wall face; its inner corner is chamfered to a height of 1.4m and is original work (Pls XI, XII). Diagonal tooling can be clearly seen on the uppermost of these chamfered stones. The north side is splayed, and its poorly finished facing and inner corner indicate secondary work, which involved cutting back of the original jamb. The upper eye stone for the door survives on this side. The rear arch is segmental and, like the external jambs, appears to be inserted. Above the doorway, roughly at eaves level, is a lintelled cupboard, now partly blocked,

and at the same level further south in this gable is another cavity in the wall, now almost completely blocked. These indicate that there may have been a loft at this end of the building at some stage.

There is a blocked-up doorway in the south wall 4.5m from the external south-west corner (Pl. XIII). No cut stones survive externally, but the narrowness and rough finish of the opening, as visible from the interior of the nave, show that it

Pl. X   The Abbey from the south-west (2002) with new roof on nave.

Pl. XI   Interior view of west doorway.

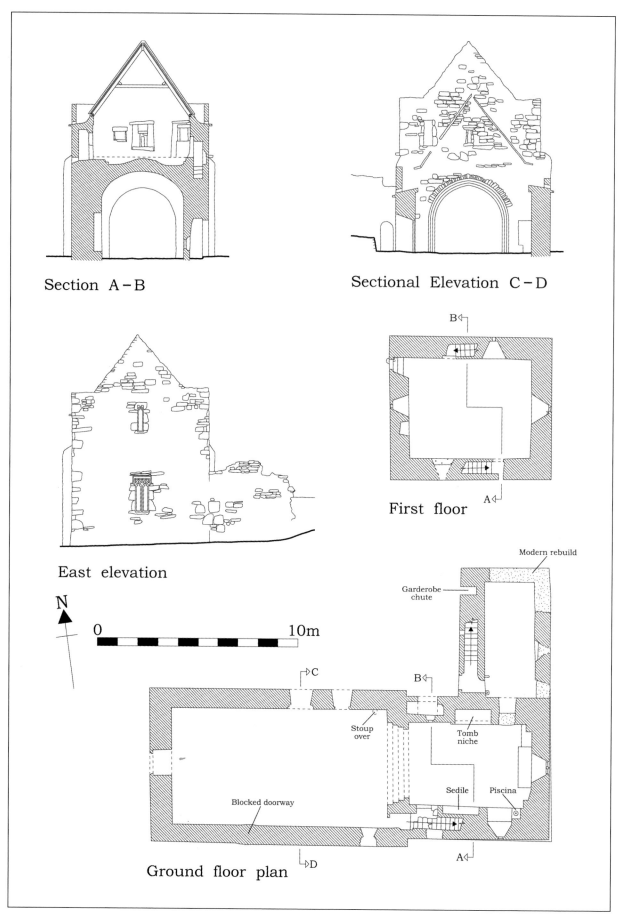

Section A – B

Sectional Elevation C – D

East elevation

First floor

Ground floor plan

N

0　　　　　　　　　　　　10m

Modern rebuild

Garderobe chute

Stoup over

Tomb niche

Sedile　　Piscina

Blocked doorway

Fig. 3　Plans, sections and elevation of the Abbey in 1992.

Pl. XII Internal south jamb of west doorway showing chamfered corner.

Pl. XIII South wall of nave showing parapet reconstructed by Office of Public Works.

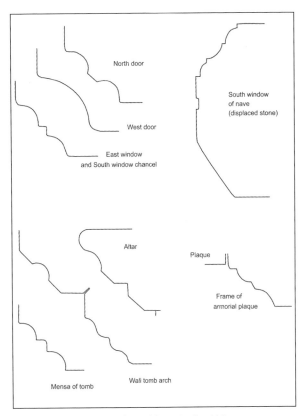

Fig. 4 Cross sections of mouldings in the Abbey.

is not an original feature. At the east end of this wall is a badly damaged window-opening with plain lintels over the internal splay and no carved stones surviving externally. One carved jamb stone presumably from this window has been reused in a masonry blocking of the lower part of the window opening, and this shows that the window was finely moulded externally and had a glass groove (Fig. 4).

The east wall has been considerably altered by the insertion of the wide chancel arch (discussed below with the chancel) and a low doorway to the stair in the south-east corner (Pls XIV, XV). However, some of the original external face of this wall can be seen in the tiny chamber in the west end of the north chancel wall, and the line of the original gable can be seen below the later roof line. The chancel arch is placed slightly north of centre in the east wall of the nave, possibly to accommodate the doorway to the stair on the south side.

There are two doorways in the north wall close together in the eastern half of the nave (Pls XVI, XVII). They are both inserted features and were not designed as external doorways but rather to connect with an attached structure (or structures), no trace of which survives. The more westerly of the two is missing its carved jamb stones, but a spud stone at the base shows evidence of use. The other doorway is complete, with a low segmental pointed head internally and a moulded chamfer

Pl. XIV   East wall of nave in 1952 showing chancel arch prior to excavation to floor levels (DOEHLG).

Pl. XV   East wall of nave during conservation of wall paintings.

Pl. XVI   Exterior of north wall of nave showing contrast between thirteenth-century wall and fifteenth-century parapet.

Pl. XVII   View from the interior of the easternmost of the north doorways in the nave (DOEHLG).

Pl. XVIII   Stoup with animal ornament set into north wall of nave.

(Pl. XVII; Fig. 4). Externally these openings do not have proper heads but were left just as they were broken through the wall (Pl. XVI). Between these doorways and the north-east corner a small projecting carved stoup has been set into the wall, probably by the Office of Public Works *c.* 1880. This stoup is 0.21m wide and roughly square in plan, with a circular cylindrical bowl 0.15m in diameter and 0.12m deep. It is made of serpentinite and has a narrower projecting tang that has been built into the wall. The outer face is carved in flat relief with two quadrepeds back to back, with their heads at the upper corners and their back feet at the lower corners (Pl. XVIII). Around these corners, on the side faces, are mirror images of each of these quadrupeds: at each corner the head of a quadruped on the outer face meets that of a quadruped on a side face (Morton, this volume, pl. XVII, p. 108). The uppermost portions of the north and south walls, with their slight inner projections and remains of parapets, and water outlets are secondary and built of much smaller stones than the walls below (Pl. XVI).

The chancel (Fig. 3)

The chancel is a one-period two-storey structure. The building stone used is generally smaller and has more of a quarried look about it than that used in the nave (Pls VIII, IX, XIX). The corner stones are mostly thin dressed stones set on edge and facing in alternate directions from the corner, in contrast to the blockier stones used in the nave. In this case the parapets and gargoyles are contemporary with the rest of the building. The east window on the ground floor has twin ogee heads and a hood moulding externally (Pl. XX). The south window on the ground floor (Pl. XXI) and the east window on the first floor are single ogee-headed lights without hood mouldings. The exteriors of the jambs are also moulded (Fig. 4). The spandrels of the window heads all have sunken pointed trefoils (Pls XX, XXI). All other windows are either partly robbed or are simpler, more roughly finished openings. There was a bell cot in the apex of the east gable prior to the reroofing of 1952. It appears to have been a late

Pl. XIX   The chancel from the south-east.

Pl. XX   Exterior view of the east window (DOEHLG).

Pl. XXI   South window of chancel (DOEHLG).

Pl. XXII   Carved head set sideways on jamb of sedile (DOEHLG).

feature, clearly inserted after the building lost its roof, and is possibly contemporary with the nearby chapel (Pl. I). The lower part of the east wall from the level of the main window down has a slight batter.

Internally the ground floor of the chancel is roughly 5.8m long and 4.03m wide. The outer order of the bluntly-pointed chancel arch is almost the same width as the chancel, and the inner order is 3.3m wide. The arch has four plain orders that follow through from the piers without any base, capital or impost. The outer order of the arch facing the nave is formed of roughly-cut voussoirs, but the remainder of the orders (both the arches and the piers) is formed of well-cut pieces of serpentinite or soapstone, which was probably quarried at one of the points where it outcrops on the island, possibly at Portruckagh, near the south-east corner of the island (Graham 2001, 5). The majority of the stones that form the orders of the arch are unusual in that their two longer faces—the surfaces forming the soffit (intrados) of the arch and those facing into the wall (the extrados)—are curved. The stones average 0.11m thick (the depth of the order). There is a rectangular slot at the springing of the outer order of the arch at each side, indicating the former existence of a wooden rood screen within the arch (Westropp 1911, pl. V).

The chancel is covered by a barrel vault and lit only by a two-light east window and a single-light window set in a deep embrasure in the south wall. The sedile, which is roughly in the centre of the south wall, is a plain pointed-arched recess 1.84m wide and 0.43m deep with a plain impost on the east side only (Fig. 3). The jambs, arch and back are of rough masonry, which was heavily plastered over and painted. Only remnants of the plaster survive. To the right of the sedile is a tall opening giving awkward access to the stair. This appears to be secondary, but its head, which has been quarried through the wall, has been plastered and finely painted. Between this doorway and the sedile there is a small stone head protruding from the wall, and there is another small carved head flush with the wall immediately above the impost on the east side (Pl. XXII). To the left of the sedile is the deep window embrasure. The jambs of the embrasure are initially at right angles to the wall but at the outer end they splay unevenly to the off-centre window. The head of the embrasure is pointed and has wattle impressions in its soffit. There is a piscina in the inner eastern corner of this embrasure, but only one side of the cusped pointed head survives. An old photograph dating from the 1950s in the National Monuments Photographic Archive shows the the other half of the head of the piscina loose on the ground (Pl. XXIII). This fragment is no longer in evidence on the site.[3]

The east window is in a large shallow arched recess almost the width and height of the chancel. The window consists of twin cusped-ogee-headed lights in a bluntly pointed splayed embrasure. The original stone altar (1.1m high) beneath it is faced with well-squared blocks of serpentinite (Morton and Oldenbourg, this volume, pl. LXIV, p. 92). The base and top project slightly and are moulded (Fig. 4), and the corner stones have an angular moulding (Westropp 1911, pl. V) that together with the upper and lower mouldings form a frame for the altar front. The altar top, formed of four pieces of serpentinite, measures 2.02m long by 0.67m wide. A solid rough masonry ledge of similar height but lesser projection than the altar runs from the altar to the north wall.

The north wall is dominated by the canopied tomb placed roughly centrally within it (Pl. XXIII, XXIV). This consists of a tomb niche (0.75m deep) with a tall ogee-headed arch with an openwork traceried screen flanked by tall upright pinnacles. In the base of the niche is a low mensa of serpentinite with a plain top. Its front panel is carved in the form of six deeply cut, cusped, ogee-headed niches: the piers continue up to form deep subtriangular spandrels. This part of the tomb is 1.98m wide and 0.56m high. All the carved work above the mensa, both pinnacles and tracery, is of limestone and, puzzlingly, is wider than the

mensa, being 2.35m wide. Plank impressions can be seen in the soffit of the pointed niche arch. The traceried screen consists of a cusped round-headed arch below an ogee-headed arch, with the apex of both connected by a perpendicular mullion. The long lentoid areas between the ogee arch and the main pointed arch of the niche are cusped at each side. For the mouldings of the canopy see Fig. 4. The hood moulding of the niche is formed into a rough ogee that ends in a pinnacle matching those on either side. Because the canopy is wider than the mensa below, extra roughly cut stones without mouldings had to be added at each end of the mensa top to support the canopy. The limestone for the canopy had to be brought from elsewhere, and if it was carved at its place of origin there is a possibility that an error was made about the width.

To the east of the tomb was, until recently, a ragged secondary doorway giving access to the north wing. This opening had weakened the structure of the canopy, causing it to bulge slightly to the east, and it has since been blocked up for structural reasons. In the east jamb of the opening, the east side of an earlier aumbry can be seen. To the west of the tomb is a seventeenth-century

Pl. XXIII   The canopied tomb as it appeared in 1952, showing the ground level before the subsequent clearance works (DOEHLG).

Pl. XXIV   The canopied tomb after the conservation of the paintings (Con Brogan, DOEHLG).

limestone armorial plaque within a moulded frame (Pl. XXV) set into the wall. The overall dimensions of this plaque, including the frame (which is formed of seperate stones) are 1.1m by 0.79m. This plaque partly blocks a narrow, rough, squint-like feature opening from the tiny wall chamber, which can be accessed from the outside through a doorway (Pl. XXVI). This chamber or cavity in the wall is a puzzling feature of uncertain purpose. It appears to have been a doorway originally, narrowing from a width of 1m externally to 0.78m close to the inner face. The lintelled upper part of the doorway survives except at the inner face, where the jambs were removed when the inner sides of the doorway were quarried into. Later the inner end of the damaged doorway was blocked with a thin wall containing a narrow opening or squint. This wall left cavities in each jamb of the doorway 1.35m high and 0.5m to 0.6m wide. On the west side the cavity extended as far as the original exterior face of the east nave wall; on the east side it only penetrated some 0.3m. The sides and head of this grave-shaped cavity were roughly plastered prior to the building of the inner blocking wall. The upper part of the opening in this thin wall was removed in order to insert the armorial plaque.

The chamber above the chancel (Fig. 3)

The straight stair (0.54m wide) in the south wall gives access to the floor above the vault through a narrow lintelled doorway. The room measures 6.15m long by 4.80m wide. There is a cusped ogee-headed window placed centrally in the east wall and plainer windows near the east end of the north wall and west end of the south wall. A plain opening centrally placed in the west wall looks out into the roof space of the nave. This is flanked by a cupboard recess to the south and and a low narrow doorway at the north corner that gave access to the roof and parapets of the nave. Centrally placed in the north wall is a tiny doorway giving access to a very cramped short straight flight of stairs descending to the first floor of the north wing, which is at a lower level. This is clearly a secondary feature. There is no surviving evidence for access from the interior to the roof- or parapet-level above. The upper surface of the vault was covered with concrete about 1910 by the Office of Public Works in an attempt to keep the vault dry. In the winter of 1952 the first floor of the

Pl. XXV   The late seventeenth-century Ó Máille plaque (DOEHLG).

Pl. XXVI   The external angle between the chancel and the north wing showing the doorway to the wing and the compartment in the north wall of the chancel.

chancel was roofed and the concrete removed from above the vault.

The north wing (Fig. 3)

The north wing—the most ruined part of the building—was a low, narrow, two-storey addition. Its east face continues the line of the east wall of the chancel (Pl. XXVII). The ground floor, which appears to have served as a sacristy, measures 5.75m long by 2.55m wide internally. Much of its north wall and the north end of its east wall is modern repair. It had a window in the east wall; part of the splayed embrasure can be seen on the inside. Apart from the doorway (now blocked) leading from the chancel there is also an entrance from the outside at the south end of the west wall. An exterior doorway here leads directly to a tiny lobby in the thickness of the wall that gives access to a narrow straight mural stair (0.56m wide) in the west wall leading to the first floor: the lobby also leads to a door straight ahead leading into the ground floor. The two phases in the latter doorway involve a reduction in the height and width in the later phase in reaction to a crack in the original lintel. The original doorway was 1.6m high by 1.06m wide, while the replacement is only 1.15m high and 0.57m wide with a rough segmental head. The eye and spud stone indicate that the door was hung on the south side; there is a projecting stone on the north side with a hole for the bolt. The ground floor was vaulted over in stone with a very flat arch, the north end of which has long since collapsed. Little survives above the first floor apart from the serpentinite sill of the two-light east window, which is plainly chamfered, in contrast to the more finely moulded windows in the chancel. A garderobe chute in the north end of the west wall indicates that a garderobe existed at first-floor level. There are also remains of an embrasure at the south end of the west wall above the doorway.

Analysis and phasing (Fig. 5)
Phase 1

The nave belongs to Phase 1 and appears to have been a single-cell rectangular church with a simple centrally placed west doorway, a window at the east end of the south wall and presumably

Pl. XXVII View from the east of the cross-inscribed pillar, chancel and north wing.

an east window. The original west doorway must have had straight jambs at right angles to the wall faces and, given that the inner part of the south jamb is original, must have been 0.85m wide to have been central in the gable wall. The corners of the doorway were chamfered and the head possibly lintelled like the transitional west doorway of Mainistir Chiaráin on Inishmore, Aran. As in the latter case, the door would have been hung on iron fittings set in the inner face of the wall to one side of the doorway. The building probably had corbels at the exterior corners to support the end rafters of a roof carried over the gables and would not have had parapets. Such a building would be likely to date from the early thirteenth century like Mainistir Chiaráin, which is of very similar dimensions (Gosling 1993, 107).

Phase 1a: In this phase a doorway was opened in the south wall of the nave. A secondary doorway in a side wall can be seen at other churches with west doorways of a similar or earlier date, such as Mainistir Chiaráin on Inishmore and St Mary's Church, Trinity Church and the Cathedral at Glendalough.

Phase 2

The nave was much altered in this phase, and a two-storey chancel was added. The south doorway of the nave was blocked and the original west doorway widened on the north side so that a door could be hung within it behind the new exterior jambs and pointed head (see Appendix, p. 26). A new south window with external mouldings (Fig. 4) was inserted in place of the original, and two doorways were broken through the north wall that were intended to give access to an attached structure or structures. It is possible that these doorways were not in use at the same time and that the easternmost one replaced the other after some time had elapsed. Either way they are puzzling because there is no visible trace of an attached stone building, and the parapet and water outlets above the doors show no change to accommodate an attached structure. The parapets and water outlets were added as part of this phase. It is possible that the intended attached structure was never built.

The chancel was a two-storey structure from the start, but evidence was uncovered during the

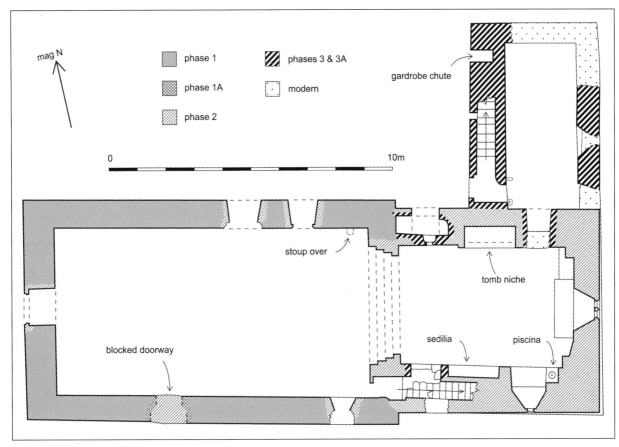

Fig. 5   Plan of the Abbey showing building phases.

recent conservation work on the paintings that the present vault is secondary.[4] The original Phase 2 building must have had either another vault that had to be replaced or a wooden first floor. The stair to the first floor appears to have been entered originally through a rudimentary doorway in the south-east corner of the nave. The canopied tomb in the north wall is an original feature but there was no doorway originally on its east side, only a small aumbry. There was an original doorway on its west side designed to give access to a planned attached structure. The interior of the chancel was painted at this stage, and parts of this original scheme of painting survive, mostly on the north and south walls, sometimes covered by a secondary layer of plaster contemporary with the present vault.

There are other comparable examples of a nave and chancel church with a room over the vaulted chancel. Possibly the closest parallel is the larger church at Liathmore, Co. Tipperary, which even has a straight stair in its south wall giving access to the first floor. In addition, the first floor was served by a garderobe on its north side (Leask and Macalister 1946, 9–12). Other examples of this

arrangement are the churches at Aghaviller, Co. Kilkenny, which also had a straight stair, in this case in the north wall, St Mullin's, Co. Carlow, which had a spiral stair, and Oughterard, Co. Kildare, which has an attached spiral stair turret.

The most ornate piece of stone work in the chancel is the canopied tomb. This should offer the best potential for dating Phase 2 but, unfortunately, this is a type that has seen little or no study. In his article, Westropp captioned it as an Easter Sepulchre (1911, pl. III), whereas in the text he suggested that it served as as a founder's tomb and an Easter Sepulchre (1911, 30). He inclined towards the latter identification because of the remains of a painted crucifixion on the rear wall of the niche. An Easter Sepulchre is a decorative canopied feature or niche in the north wall of the chancel of some English churches. It was used in the Easter ceremonies. While it was probably originally a wooden feature, examples in stone are found from the fourteenth century on (Anderson 1971, 158, pl. 67). Some Easter Sepulchres in England doubled up as tombs (Duffy 1992, 32), but until further research is done on these monuments in Ireland the question as to whether the tomb on Clare Island also served as an Easter Sepulchre must remain open.

The tomb niche as a type appears to have been introduced to Ireland with the Gothic style in the thirteenth century, and examples can be seen at Gowran Church and St Canice's Cathedral in Kilkenny and at Ardmore in County Waterford (Leask 1960a, 143–4). These often have trefoil or cinquefoil arches: a row of them, forming an arcade, is sometimes found, as at Gowran. Other thirteenth-century examples have projecting gables over the arch, as at St Mary's church in New Ross, Co. Wexford, and Corcomroe, Co. Clare, but none of these early examples has tracery. What is possibly the earliest example with tracery is the tomb at Dungiven priory, Co. Derry. This was for a long time accepted as the tomb of Cooey na Gall O'Cahan, who died in 1385, until Hunt (1974, vol. I, 130–2) dated it to the last quarter of the fifteenth century. Hunt's dating has recently been challenged by McNeill (2001, 348–52), who has made a strong case in support of the original identification of the tomb as that of Cooey na Gall and consequently a late fourteenth-century date. The style of the tomb and tracery is very different from that of the Clare Island tomb, but the statement being made is similar: that these were aristocratic burial places in a common European tradition. The perpendicular Gothic style exhibited in the Clare Island tomb remained popular from the late fourteenth century until the mid-sixteenth century. Possibly one of the earliest tombs in this style in Ireland is the Galwey-Bultingfort tomb in Limerick cathedral, dating from c. 1400 (Leask 1960b, 170–2). This tomb does not have tracery. Tomb niches with tracery (like the Clare Island example) are a feature of abbeys and friaries, particularly in the west of Ireland, and date mainly from the fifteenth and early sixteenth centuries (Leask 1960b, 167–74). These tombs merit further study to refine their dating but the example here is likely to be of fifteenth-century date and could belong to the middle or earlier half of the century. There is a need also for further study on mouldings, and when such work has been done it may be possible to arrive at a closer dating for Phase 2 on the basis of the recorded mouldings (Fig. 4). The possibility that Diarmait Ó Máille, who became chief in 1415, built the chancel is supporting evidence for an early fifteenth-century date for this phase.

Phase 3

The present vault of the chancel (contemporary with the second phase of painting) and the added north wing may belong to the same phase because the doorway that was broken through the north wall to the sacristy in the wing cut through the first phase of painting. The blocking of the doorway in the west end of the north wall of the chancel and the formation of a cavity in the wall with a squint looking into the chancel is probably also part of this phase of work. The stair connecting the first floors of the chancel and north wing was partly quarried through the wall at this stage, although it looks as if it made partial use of some pre-existing feature. The greater and best known part of the painting, that on the vault itself, dates from this phase. There is nothing in the architecture of this phase that is closely datable but, assuming some time lapse between it and Phase 2, it is likely to date from at least the middle decades of the fifteenth century. A radiocarbon determination from wattle fragments from the vault of the chancel gave a date-range of 1270 to 1420 AD at 95% certainty. This would suggest an earlier date for Phase 3, but it is dangerous to place too much reliance on one radiocarbon sample and there is also a 5% chance of the actual date being outside the given range.

Later work

The inner west doorway of the north wing was altered subsequent to Phase 3, and the stair here appears to have been roughly widened by chiselling away parts of its inner faces, possibly to accommodate a corpulent cleric. The O'Malley armorial plaque dates from the seventeenth century and must have been inserted into the north wall of the chancel at that time. Another very late feature was the bell cot, added to the east gable of the chancel after it had become unroofed. It was removed when the chancel was reroofed in 1952.

Conclusion

The surviving remains of the Abbey agree in general terms with the very scanty historical record of the place. The nave was a simple rectangular church of about the early thirteenth century, around the time the foundation was annexed to the Cistercian abbey of Knockmoy. If the church was built by or for any sort of monastic community, it could be seen as conforming to the older Irish monastic tradition, where domestic structures were not directly attached to the church and were built of perishable materials. It certainly conformed to the older tradition of church building in stone in being a simple rectangle in plan, with a west doorway and very limited number of windows, one in the east wall (removed when the later chancel arch was built) and one in the south wall. It remains uncertain whether this church was built by or for the Cistercians or was already there when the place was given to the Cistercians, or whether there ever was an intention of founding a full abbey here.

Little change can be detected in the structure in the thirteenth and fourteenth centuries apart from the opening of a doorway in the south wall, probably necessitated by some problem with the west doorway, such as a crack in its lintel. Presumably the west doorway was blocked up when the south doorway was opened.

Major additions were made to the building in two phases in the fifteenth century. In the first of these a new chancel with a room above it was added to the east end of the older church and the walls and presumably the ceiling (whether stone or wood) were painted. In a second fifteenth-century phase, possibly not long after the first, a new vault was inserted to form the ceiling and and the walls were newly painted. Probably also as part of this phase a domestic north wing was added to the chancel.

The fifteenth-century building campaigns are likely to have been sponsored by the local Gaelic lords, the O'Malleys, partly to provide a burial place for themselves. There is corroboration that the O'Malleys were patrons of the Abbey in a late seventeenth-century source. The accommodation above the chancel and in the north wing was presumably for the priest, monk or monks who were attached to the church. Doorways in the north wall of the nave indicate that there may have been more ambitious plans for attached domestic structures that were never realised.

The fifteenth century was the time when there was a flowering of religious foundations, especially friaries, in the west of Ireland, and the probable O'Malley patrons of the Abbey may later have been associated with the foundation of an Augustinian friary at Murrisk, under Croagh Patrick. At this time the Abbey could be seen as the equivalent of smaller religious establishments in Gaelic Ireland, such as those on islands, some of which had adopted the Augustinian rule and some of which appear to have continued in the older Irish monastic tradition. Examples of the latter were on Inchcleraun and Inishb-offin on Lough Ree. The provision of domestic accommodation in or attached to the church was common in the fifteenth and sixteenth centuries all over Ireland. Mostly this accommodation was at the west end of the church or in an attached residential tower, again usually at the west end.

The provision of accommodation above the chancel is not so common and a number of parallels have already been listed. Another example that could be added is Clonmacnoise cathedral, where a room above the vaulted east end was added in the 1450s (Manning 1998). The number of these examples where there was also an early medieval round tower on the site is surprising (Clonmacnoise, Oughterard, Liathmore, St Mullin's and Aghaviller), but it is not clear what the presence of both implies. It appears that at some ecclesiastical sites that were important in early medieval times, it was appropriate in later medieval times to provide accommodation for the resident priest or monk above the chancel of the church. In other cases it may have been the monastic affiliations of certain churches (such as the Abbey on Clare Island, St Mullin's, Co. Carlow, and Dungarvan, Co. Kilkenny) that made it appropriate to have accommodation above the chancel.

The Abbey is an interesting survival, and the extent of its mural paintings makes it unique in an Irish context. This is all the more surprising considering its location on an island exposed to the Atlantic.

Acknowledgements

I owe a particular debt of gratitude to David Sweetman and Brian Ronayne, who carried out a detailed survey of the building in July 1993 as part of the New Survey of Clare Island. The plan, section and elevations were drawn up by Brian Ronayne. The building phases, for which I alone am responsible, were subsequently added to the plan by Tom O'Sullivan. I am also indebted to Gerard Woods for the final versions of Figs 1, 2 and 3, and to Con Brogan, John Scarry and Tony Roche for assistance with the photographs. For discussion on the Abbey over many years I am grateful to Roger Stalley, Christoph Oldenbourg, Karena Morton, Paul Gosling, Willie Cumming and David Sweetman. I am grateful to Dr John Graham for identifying the serpentinite, used so extensively for the finer stonework, and to the late Liam de Paor for translating the piece from Jongelinus.

APPENDIX
Supervision of excavation of drainage trenches at the Abbey in 1991

In an attempt to dry out the building in connection with the conservation work on the paintings, drains were inserted within the church and along the outside of the north and west walls in September 1991. The author supervised the excavation involved.

In the chancel, trenches (0.30m wide and 0.20m deep) were dug close to all four sides and plastic drainage pipes were inserted. Nothing was found apart from disturbed human bones, which were reburied on site. A single trench was dug down the centre of the nave, and in order to drain the trenches in the chancel westwards, its depth was increased gradually from 0.25m at the east to 0.40m at the west. It was cut through disturbed soil that contained only a few small fragments of human bone.

The west doorway was excavated, and we found it was possible to accommodate the drainage pipe below the threshold stones. The outer threshold stone clearly belonged to Phase 2 and was at a high level (Pl. XXVIII). At the inner side of the doorway was a larger threshold stone at a lower level, which projected under both jambs. It must belong to Phase 1. No spud stone for the door survived.

Outside the nave on the north side the trench ran from the angle between the chancel and the north wing to the north-west corner of the nave and gradually increased in depth from 0.15m at its eastern end to 0.30m at the west. The only place

Pl. XXVIII   The west doorway of the nave during excavation 1991.

where occupation deposits were encountered was between the westernmost north doorway and the north-west corner of the nave. These deposits contained shells (mostly limpets but with some cockles and periwinkles) and a few scraps of animal bone. There was also some heat-reddened clay. No artefacts were recovered to indicate a date for these deposits. No traces of wall foundations for attached buildings that could be entered through the north doorways were encountered in this area.

At its west end the drain was kept close to the wall, which had a foundation plinth projecting 0.05–0.10m. A sump hole was dug in a clear patch 1m south of the south-west corner of the nave.

NOTES

1. Luckombe (*A tour through Ireland*, 1780, 216) was in fact referring to the friary on Sherkin Island, Co. Cork.
2. This Latin word means 'lame' and is therefore the equivalent of *bacach* in Irish.
3. There was a tradition of building a stone from the Abbey into any new house that was being built on the island (Críostóir Mac Cáirthaigh, pers. comm.). This may explain the loss of this stone.
4. This conclusion is based entirely on the observations of Christoph Oldenbourg during the conservation work, in that he saw evidence of original plaster on the north and south walls continuing vertically upwards behind the plaster on the vault. He is also of the opinion that the pilaster-like features in the south-east and north-east corners of the chancel and the arch above them are Phase 3 additions. The present writer is not convinced of this, and this element of the building is not shown as Phase 3 on Fig. 5.

REFERENCES

Alemand, L.A. 1690 *Histoire monastique d'Irlande*. Paris. Michel Guerout.

Alemand, L.A. 1722 *Monasticon Hibernicum, or the monastical history of Ireland*. J. Stevens (trans.). London. William Mears.

Anderson, M.D. 1971 *History and imagery in British churches*. Edinburgh. John Murray.

Deane, T.N. 1880 Clare Island. In appendix to *The 48th report of the Commissioners of Public Works in Ireland*, 75. HMSO.

Duffy, E. 1992 *The stripping of the altars: traditional religion in England c. 1400–c. 1580*. New Haven and London. Yale University Press.

Freeman, A.M. (ed.) 1970 *Annála Connacht: the Annals of Connacht (A.D. 1224–1544)*. Dublin. The Dublin Institute for Advanced Studies.

Gosling. P. 1993 *Archaeological inventory of County Galway. Volume 1: west Galway*. Dublin. Stationery Office.

Graham, J.R. 2001 The geology of Clare Island: perspectives and problems. In J.R. Graham (ed.), *New survey of Clare Island. Volume 2: geology*. Dublin. Royal Irish Academy.

Gwynn, A. and Hadcock, R.N. 1970 *Medieval religious houses, Ireland*. London. Longman.

Hayes-McCoy, G.A. 1996 *Scots mercenary forces in Ireland (1565–1603)*. 2nd edn. Dublin. Edmund Burke Publisher.

Hunt, J. 1974 *Irish medieval figure sculpture 1200–1600*, 2 vols. Dublin and London. Irish University Press.

Jongelinus, G. 1640 *Notitiae abbatiarum ordinis Cistertiencis per universum orbem*. Coloniae Agrippinae. Apud Ioannem Henningium Bibliopolam.

Leask, H.G. 1960a *Irish churches and monastic buildings. Volume 2: Gothic architecture to A.D. 1400*. Dundalk. Dundalgan Press.

Leask, H.G. 1960b *Irish churches and monastic buildings. Volume 3: medieval Gothic, the last phases*. Dundalk. Dundalgan Press.

Leask, H.G. and Macalister, R.A.S. 1946 Liathmore-Mochoemóg (Leigh), County Tipperary. *Proceedings of the Royal Irish Academy* **51**C, 1–14.

Luckombe, P. 1780 *A tour through Ireland*. J. Place and R. Byrn for Mssrs Whitestone, Sleater, Sheppard.

McNeill, T. 2001 The archaeology of Gaelic lordship east and west of the Foyle. In P.J. Duffy, D. Edwards and E. Fitzpatrick (eds), *Gaelic Ireland: land, lordship and settlement c. 1250–c. 1650*, 346–56. Dublin. Four Courts Press.

Manning, C. 1998 Clonmacnoise Cathedral. In H. King (ed.), *Clonmacnoise studies vol. I: seminar papers 1994*. Dublin. Dúchas.

Murphy, D. (ed.) 1891 *Triumphalia chronologica Monasterii Sanctae Crucis in Hibernia*. Dublin. Sealy, Bryers & Walker.

Ó Muraíle, N. 1998 A description of County Mayo c. 1684 by R. Downing. In T. Barnard, D. Ó Cróinín and K. Simms (eds), *'A miracle of learning': studies in manuscripts and Irish learning. Essays in honour of William O'Sullivan*, 236–65. Aldershot. Ashgate.

Ó Muraíle, N. 1999 The place-names of Clare Island. In C. Mac Cárthaigh and K. Whelan (eds), *New Survey of Clare Island. Volume 1: history and cultural landscape*, 99–141. Dublin. Royal Irish Academy.

O'Sullivan, W. (ed.) 1958 *The Strafford inquisition of County Mayo*. Dublin. Stationery Office.

O'Sullivan, W. 1997 A finding list of Sir James Ware's manuscripts. *Proceedings of the Royal Irish Academy* **97**C, 69–99.

Otway, C. 1839 *A tour of Connaught: comprising sketches of Clonmacnoise, Joyce Country, and Achill*. Dublin. William Curry, Jun. and Company.

Simington, R.C. (ed.) 1956 *Books of survey and distribution. Volume 2: county of Mayo*. Dublin. Stationery Office.

Stalley, R. 1987 *The Cistercian monasteries of Ireland*. London and New Haven. Yale University Press.

Ware, J. 1626 *Archiepiscoporum Casseliensium et Tuamensium vitae; duobus expressae commentariolis. Quibus adjicitur historia coenobiorum Cisterciensium Hiberniae*. Dublin. Ex Officina Societatis Bibliopolarum.

Ware, J. 1658 *De Hibernia et antiquitatibus eius disquisitiones*. 2nd edn. London. Jo. Crook.

Westropp, T.J. 1911 Clare Island Survey: history and archaeology. *Proceedings of the Royal Irish Academy* **31** (1911–15), section 1, part 2, 1–78.

THE ABBEY AND ITS ASSOCIATED ARCHAEOLOGICAL REMAINS

Paul Gosling

ABSTRACT

The complex of religious monuments associated with the Abbey is described and discussed, showing that this site was and continues to be the main ecclesiastical centre on the island, with a tradition stretching back to early medieval times.

Introduction

The Abbey at Kill is undoubtedly the jewel in the crown of Clare Island's built heritage. Its association with the Cistercians has long been recognised and has afforded the site a certain status in the academic literature (e.g. Westropp 1911, 29–37; Stalley 1987, 123–5, 243). However, the complexity of its architectural history has only recently become apparent. When coupled with the degree of preservation of its internal fittings, especially its murals, it must now be ranked as an important building in Ireland's canon of medieval architecture. The Abbey's physical remains also serve to indicate its status on a more local scale. As a central place in the island's historic topography, the Abbey's only rival is the natural harbour at Glen, which lies 2km to the east. There the ruin of the castle indicates the seat of temporal power on the island in late medieval times. Even today, a clear majority of Clare Island's communal services are located at Kill. These include the Roman Catholic Church, the National School, the post office, the shop and the graveyard.

This serves to emphasise that the Abbey is not an isolated physical feature. Rather, it is a symbolic marker for the island's cultural centre, specifically as its most important religious site. In this latter role, the Abbey is simply the most visible of a cluster of archaeological and historical remains of varying age. These include the site of a Roman Catholic chapel (no. 163),[1] a cross-inscribed pillar stone (No. 147), a holy well (No. 151), an 'altar' (no. 152) and a possible bullaun (a stone with a bowl-shaped depression) (no. 150), now lost. While the well and altar are associated with a separate enclosure (no. 78) 125m north-west of the Abbey, the other remains are all within the confines of the graveyard (no. 146).

The graveyard

The immediate frame of reference for the Abbey is the graveyard itself. This sits just above the thirty metre contour and overlooks the seashore some 250m to south. Rectilinear in plan (54m long and a maximum of 31.5m wide), its long axis runs east–west along a natural south-facing terrace. In its present form, the graveyard does not appear to have altered greatly since it was first mapped by the Ordnance Survey in 1838 (Pl. Ia). However, its western boundary has disappeared in the meantime, probably when the adjoining Roman Catholic Church was built in 1861.[2] Furthermore, the present boundary walls probably date only from *c.* 1890 (Pl. Ib). This is revealed by two brief but contrasting reports made in 1888 and 1891 that are published in the *Journal of the Association for the Preservation of the Memorials of the Dead in Ireland*. The first report (Vigors 1890, 24) decries the disgraceful state of 'Grace O'Malley's tomb', but the second (Vigors 1893, 454–5) notes that 'the

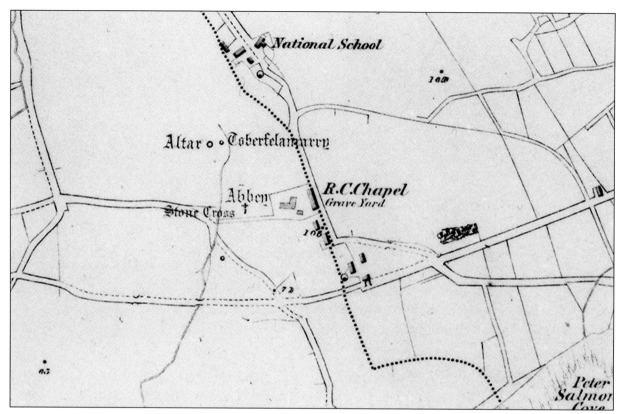

Pl. Ia The topography of Kill, Clare Island, as shown on the first edition of the OS six-inch scale maps. Apart from the ecclesiastical remains, note also the forge (indicated by Ω) and the two lime kilns (indicated by ☉). Extract from County Mayo, Sheet 85 (Ordnance Survey of Ireland 1840).

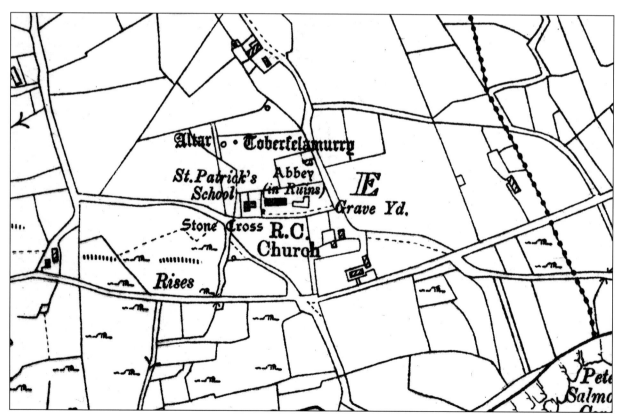

Pl. Ib The topography of Kill, Clare Island, as shown on the third edition of the OS six-inch scale maps. Note the change in the townland boundary (the dotted line running roughly north–south) between 1840 and 1920. Extract from County Mayo, Sheet 85 (Ordnance Survey of Ireland 1920).

abbey and graveyard are now protected by a good wall', temporising that 'SKULLS and bones are still lying about, exposed (!)'.

All this contrasts with the neat appearance of the graveyard today, with its mown grass and well-maintained headstones. Some 150 grave plots are visible, most dating from the later nineteenth and twentieth centuries (see Cantwell and Gosling, this volume, pp 35–40). The spatial and temporal distribution of these graves is notable (see Cantwell and Gosling, this volume, fig. 1, p. 37). In particular, there are relatively few gravestones along the whole gently sloping northern part of the graveyard. Furthermore, all but one of the pre-1900 AD gravestones are to the south and east of the medieval Abbey. The specific cultural or physical forces that dictated this pattern are unclear. It could be an expression of the general tendency in pre-twentieth century burial placement in Ireland of avoiding interments in the areas to the north of a church. However, local tradition labels the ruined northern wing of the Abbey buildings as 'clauster' (cloister), while the area to the north of it is known as 'Teach na mBan'. This raises the possibility that the grave placement is avoiding the buried remains of other buildings.

Though the Abbey buildings are the only substantial remains now visible within the graveyard, the first edition of the Ordnance Survey six-inch map shows two other structures (Pl. Ia). Firstly, a tiny, rectangular, roofed structure (c. 7m × c. 5m) is shown just to the south-east of the Abbey. Its long axis ran east–west, but no physical surface trace or memory of it now survives: it may have been a vault or memorial. The second was a chapel (no. 163), situated on the eastern boundary of the graveyard. It appears on the Ordnance Survey map as a long narrow building (c. 22m × c. 8m), its long axis running roughly north–south. Its designation as 'R.C. Chapel', when combined with its representation on the 1838 map as a roofed building, suggests that it was in use at the time. Lewis (1837, vol. 1, 336) mentions that 'in the R.C. division the islands of Clare and Innisturk form a parish, in which are places of worship, but no regular chapel' (see also O'Flanagan 1927, 418). Though no surface traces of this chapel now survive, the foundation of a mortared stone wall running north–south was uncovered in the south-eastern corner of the

graveyard in 1990. Revealed during the digging of a grave (Cantwell and Gosling, this volume, grave plot no. 12, p. 39), it sparked a comment from one islander, John Lizzie O'Malley (then aged c. 92 years) that it was part of 'the old school'.[3] The present school dates from the 1880s: a stone plaque at the schoolyard gate reads 'St Patrick's N.S. 1887'. It replaced an earlier 'National School', which stood a short distance to north of the graveyard (see Pl. Ia). Perhaps the old chapel was used as a school for a time?

The cross-inscribed pillar stone

The most visually striking feature in the graveyard is the cross-inscribed pillar stone (no. 147; Fig. 1, Pls II, III). This now stands 16m east of the 'Abbey', but on the six-inch Ordnance Survey map of 1838 it appears c. 45m to the west of it, well outside the graveyard's limits (Pls Ia, II). When exactly it was moved is unclear (see catalogue entry no. 147, Vol. 5), but it was certainly in its present position by the 1950s. Standing 2.65m in height, the pillar stone is rectangular in section and profile save for its distinctively sloped top, which has a small socket-like depression. It bears a simple Latin cross on its eastern face, the arms of the cross springing from the shaft three-quarters of the way up. The top of the cross is squared off by a horizontal line 0.35m below the top of the pillar, but the ends of the arms simply terminate at the pillar edges.

In terms of its scale, the pillar stone at Kill is not without parallel. For instance, it may be compared to the slender cross-inscribed pillar at Doonfeeny graveyard, near Ballycastle, Co. Mayo, which is just over 5m in height. Parallels for the socket at the top can also be cited: an example, albeit on a smaller scale, appears on a diminutive stone cross from Templemoyle graveyard, near Malin Head, Co. Donegal (Lacy 1983, fig. 128b).

Determining the date and function of the pillar stone is much more difficult. In the first instance, one cannot be absolutely certain that pillar and cross are coeval. Cullen and Gill (1992, 2–3) have focused on this, suggesting that the cross was 'probably cut in a standing stone already in situ'. While this may have been the case at Doonfeeny it seems unlikely here, given the precise way in which the cross is carved to match the dimensions of the pillar. However, such a suggestion cannot

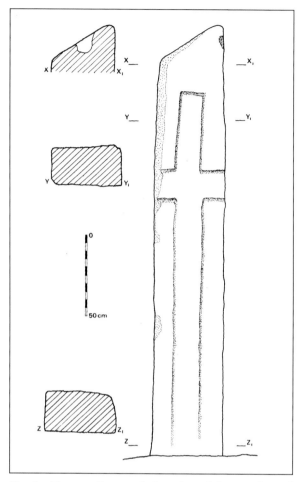

Fig. 1 Plan, sections and elevation of the east face of the cross-inscribed pillar stone (no. 147) in the graveyard at Kill, Clare Island.

Pl. III The east face of the cross-inscribed pillar stone (no. 147) at Kill, Clare Island.

Pl. II Old photograph showing the pillar stone in the boundary wall between the Roman Catholic Church and the school to the west of the Abbey (DOEHLG).

be dismissed lightly, as the lines forming the shaft do, in fact, peter out at the pillar's foot, *c.* 0.1m above the sod. Secondly, the style of the cross itself—a plain two-line, equal-armed Latin cross—is notoriously difficult to date. This is exemplified in Jim Higgins' study of Galway cross-slabs and pillar stones (1987, 50–2, fig. 17). He includes such crosses under his 'Type II' category, which he dates to the seventh, eighth and ninth centuries (Higgins 1987, 178). However, he also acknowledges that crosses of similar style 'were still being carved throughout the 18th and 19th centuries' (Higgins 1987, 179–80), noting that 'their simplicity of form, and their plainness makes it virtually impossible to distinguish between early and late examples' (cf. Lacy 1983, 279–80, fig. 150a). Given such latitude, any certainty regarding dating is impossible. However, when the style of the Clare Island cross is viewed in the context of the scale of the pillar stone and its topographical setting, a date sometime between the seventh and twelfth centuries AD does seem likely. As to the purpose of the pillar stone at Kill, the cross may have functioned as a termon or boundary marker, marking the approaches to the site from the west. As such, it can be readily compared to any of the four cross-inscribed pillar stones of similar dimensions at Mainistir Chiarán, on Inishmore, in the Aran Islands (Waddell 1994, 112–13).

The holy well and altar

Two other ecclesiastical features in the vicinity of the Abbey also deserve mention. These are a holy well and an altar (nos 151 and 152), both of which lie within a small enclosure (no. 78) some 125m to the north-west of the graveyard (Pl. I). Though neither of them are particularly well-preserved, their presence close to the Abbey is notable. The well is labelled on the Ordnance Survey six-inch map as 'Toberfelamurry', indicating its dedication to the feast of the Blessed Virgin Mary. This is generally celebrated on the 15th of August, and the name suggests that this well once witnessed gatherings on that day. In their present form, the altar and enclosure may well be relics of such ritual observances, as at Capnagower at the eastern tip of Clare Island. There, the conjunction of a holy well, altar and enclosure (nos 73 and 141–3) acted as a focus for an annual pattern (patron day celebration) within living memory. However, Westropp (1911, 28–9) makes no reference to a pattern at Kill at the time of his visit and no local memory of such survives.

Conclusion

Though the Abbey and its associated monuments are generally referred to as being at Kill, they all now lie well within the boundaries of the townland of Strake (Pl. Ib). However, this state of affairs is of relatively recent vintage, the result of major boundary changes ushered in by the Congested Districts Board in the late 1890s (see Whelan 1999, maps 1 and 2). As the first edition of the Ordnance Survey six-inch maps shows (Pl. Ia), the graveyard was originally situated in Lecarrow townland. In fact, the boundary between it and the adjoining townland of Kill ran along the eastern wall of the graveyard. The foregoing raises the question of the antiquity and origins of the ecclesiastical activity at Kill. The spatial associations of townland boundaries and medieval ecclesiastical remains is a recurrent phenomenon in Ireland (Swan 1983, 276). Though the precise significance of such associations is little understood, it is regarded by Swan (1983, 274) as one of twelve features that are characteristic of pre-twelfth century ecclesiastical sites. In fact, if we apply Swan's criteria to the ecclesiastical remains at Kill, they produce some interesting results. The spatial association of the cross-inscribed pillar, holy well, graveyard and townland boundary, in conjunction with the placename evidence (Kill), means that we can fulfil five out of his twelve diagnostic features. Thus, it is probable that ecclesiastical activity at Kill stretches back before 1200 AD. This is reinforced by the absence of any other potential early ecclesiastical sites on the island (cf. Westropp 1911, 25). As to the form or extent of this early ecclesiastical activity nothing is presently known apart from what has been presented above. It may have been a fully fledged monastic settlement or a hermitage but could just as easily have been a simple early Christian cemetery.

NOTES

1. The numbers accompanying each item provide the reader with a cross-reference to the detailed 'Catalogue of archaeological sites, monuments, and find spots on Clare Island', in P. Gosling, C. Manning and J. Waddell (eds), *New Survey of Clare Island. Volume 5: archaeology* (Dublin, Royal Irish Academy, forthcoming). The Abbey itself is no. 148.

2. A heart-shaped stone plaque high up on the east gable of the church reads 'The Church of the Sacred Heart of Jesus 1861'.

3. Information from Mr Bernie Winters, Kill, and Mr Marty O'Malley, Bunnamohaun.

REFERENCES

Cullen, C. and Gill, P. 1992 *Holy wells and Christian settlement on Clare Island, Co. Mayo*. Clare Island Series 4. Clare Island. Centre for Island Studies.

Gosling, P., Manning, C. and Waddell, J. (eds) forthcoming *New Survey of Clare Island. Volume 5: archaeology*. Dublin. Royal Irish Academy.

Higgins, J. 1987 *The early Christian cross slabs, pillar stones and related monuments of county Galway, Ireland*, 2 vols. BAR International Series 375. London. HMSO.

Lacy, B. 1983 *Archaeological survey of County Donegal*. Lifford. Donegal County Council.

Lewis, S. 1837 *A topographical dictionary of Ireland*, 2 vols. London. S. Lewis & Co.

O'Flanagan, M. (ed.) 1927 Ordnance survey name books Co. Mayo no. 100: from Kilcommon to Kilgeever. Unpublished typescript in Hardiman Library, National University of Ireland, MS D15.54.

Stalley, R. 1987 *The Cistercian monasteries of Ireland*. London and New Haven. Yale University Press.

Swan, L. 1983 Enclosed ecclesiastical sites and their relevance to settlement patterns of the first millennium A.D. In T. Reeves-Smyth and F. Hammond (eds), *Landscape archaeology in Ireland*, 269–94. British Series 116. Oxford. Archaeopress.

Vigors, P.D. 1890 Reports from counties: County Mayo. *Journal of the Association for the Preservation of the Memorials of the Dead, Ireland* **1**, no. 1, 24.

Vigors, P.D. 1893 Reports from counties: County Mayo. *Journal of the Association for the Preservation of the Memorials of the Dead, Ireland* **1**, no. 4, 452–7.

Waddell, J. 1994 The archaeology of Aran. In J. Waddell, J.W. O'Connell and A. Korff (eds), *The book of Aran, the Aran Islands, Co. Galway*, 75–135. Kinvara. Tir Eolas.

Westropp, T.J. 1911 Clare Island Survey: history and archaeology. *Proceedings of the Royal Irish Academy* **31** (1911–15), section 1, part 2, 1–78.

Whelan, K. 1999 Landscape and society on Clare Island 1700–1900. In C. Mac Cárthaigh and K. Whelan (eds), *New Survey of Clare Island. Volume 1: history and cultural landscape*, 73–98. Dublin. Royal Irish Academy.

THE GRAVESTONE INSCRIPTIONS OF CLARE ISLAND UP TO 1901

Ian Cantwell and Paul Gosling

ABSTRACT

This paper contains details of all the gravestone inscriptions up to and including 1901 in the Abbey and the surrounding graveyard at Kill.

Introduction

The gravestone inscriptions presented here are all located in or near the graveyard (no. 146)[1] adjoining the modern Roman Catholic church (Pl. I), which is located midway along the south side of the island. Though this area is known locally as Kill, the church and graveyard are actually located in the townland of Strake. Associated with them is a cluster of ecclesiastical remains of various dates. These comprise the Abbey (no. 148), a cross-inscribed pillar stone (no. 147), the site of the old RC chapel (no. 163), a holy well (no. 151) and an altar (no. 152).

As part of the original Survey of Clare Island, Westropp (1911, 37) made an abbreviated list of all the pre-1870 'inscribed tombstones' in the Abbey and graveyard. In the 1950s, the Office of Public Works carried out extensive conservation works at the Abbey (see Manning, this volume, pp 7–27). This included the removal of c. 0.5m depth of accumulated soil and debris from the interior of the church, resulting in the relocation of some inscribed grave stones (see 17 and 20, below). The inscriptions presented here were transcribed by Brian Cantwell at Easter 1985 (Cantwell 2002) and correlated with the archaeological survey work carried out by Paul Gosling in December 2001.

This paper contains details of all the inscribed grave memorials dated to or before 1901, the date of the first population census for which complete returns survive. Where a memorial or set of memorials has dates before and after 1901 a complete transcription was made. All undated inscribed memorials considered to be of pre-1901 date are also included. The entries are arranged alphabetically by the surname of the first person mentioned on each inscription. The numbers that prefix most entries refer to the numeration of the memorials on the detailed plan of the gravestones (Fig. 1). Where more than one surname occurs on a memorial, it is given a separate un-numbered entry and cross-referenced to the entry that contains the full inscription. All the memorials are located within the boundary walls of the graveyard.

The complete text inscribed on each gravestone has been transcribed, including abbreviated dedicatory phrases such as 'IHS' and 'RIP', as well as the names of the monumental masons. The texts are transcribed in lower case letters throughout, except in the case of the first letter of the inscription, proper names and placenames, where initial capitals are used. Where the reading of a letter or word is in doubt, it is placed in parentheses. Where one or more words are missing, a string of dots appears (an ellipses) (see 17). Some grave memorials located by Cantwell in the interior of the Abbey in 1985 could not be re-located in 1999 due to conservation work on the mural paintings, i.e. 5, 15 and 17.

The information provided on these gravestones is mainly genealogical, but they also contain some interesting topographical and social detail. Note the priest, Rev. Davies, who died in

Pl. I   Aerial view of the graveyard at Kill, Clare Island, showing the Abbey and the layout of the graves (Con Brogan, DOEHLG).

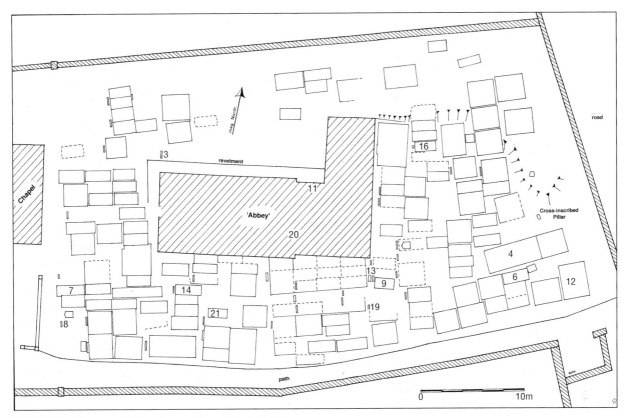

Fig. 1   Plan of graveyard showing positions of slabs 1–21.

Rome (5), the 'lightkeeper', John Gillespie (6), and the mention of the watermill at Maum (15). The earliest memorials are probably those with Latin and/or Irish inscriptions (11 and 12). Although neither is dated, that in the north wall of the Abbey chancel (11) probably dates to the late seventeenth century (see Gillespie and Ó Comáin, this volume, pp 41–44). The other Latin inscription is probably of similar date. It is carved on a rough slab recovered from a modern grave in 1990 (12). Though it begins with the phrase *Hic jacet* 'Here lies', it is not fully decipherable. It would appear to have been a rough draft for the text of an inscription, hence the horizontal and vertical guidelines (see Fig. 2).

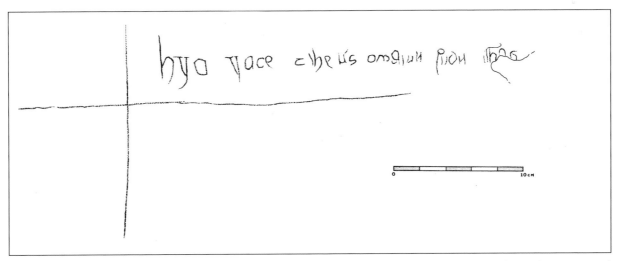

Fig. 2   Drawing of the inscription on the sandstone slab (12) recovered during the digging of a grave in the south-east corner of the graveyard.

The inscriptions

1. *[Anonymous]*

Notes: At the holy well (no. 151). Westropp (1911, 28) noted 'three dedicatory slabs' at the altar (no. 152), which lies 100m north-west of the Abbey (no. 148): two of these are now at the nearby well (no. 151; see 2 and 10 below.) The third was not located, but is illustrated by Westropp (1911, 29, pl. V).

Inscription: 'IHS B:V'

BARRET, Peter—see 19.
BARRET, Walter—see 19.

2. *[Anonymous]*

Notes: At the holy well (no. 151): see Pl. II. Noted by Westropp (1911, 28), who gives the name 'D.M. Philip'.

Pl. II   View of the two dedicatory slabs (2 on right, 10 on left) lying on the surface of the cairn that covers the Holy Well (no. 151).

Inscription: '(T?)o God this day/ dedicate to the Ble(s?)/=ed Virgin Mary/ AD 479 and (A)_1790/ (p)m: phil [*stone fractured*] /[*another illegible line*]

3. *BULGER, Thomas*

Inscription: 'IHS/ O Lord have mercy/ on the soul of Thomas/ Bulger who departed/ this life May the 9th/ 1818 aged 26 years/ Gloria in exc(e)l(s)is Deo'

4. *BYRNE, Terisa*

Description: Two adjacent memorials.

Inscription: 'Have mercy O Lord/ on the soul of/ Terisa Byrne/ who died 4th Dec 1901/ aged 30 years/ RIP/ Monaghan Westport'

Inscription: 'To the memory of McCabe family/ James J. 18-1-1944/ Cecilia 19-5-1947/ Bernard 4-5-1926/ Peter J. 7-12-1942/ Michael J. 26-6-1968'

5. *DAVIS, JAMES*

Notes: In the porch of the RC church.

Inscription: 'O Lord have mercy on the soul/ of the Rev James B. Davis who/ erected this church in 1862/ and died at Rome/ April 17th 1864 aged 34/ RIP'

6. *GILLESPIE, John*

Inscription: 'O Lord have mercy/ on the soul of/ John Gillespie/ lightkeeper/ who died 15th July 1889/ aged 30 years/ RIP/ Monaghan Westport'

7. *JOYCE, Patrick*

Notes: Two associated memorials.

Inscription: 'May the Lord/ have mercy/ on the souls of/ Patrick Joyce/ who died 15th July 1865/ aged 36 years/ and his mother/ Catherine/ who died 20th May 1876/ aged 78 years/ also John Joyce/ who died 20th August 1894/ aged 33 years/ may they rest in peace'

Inscription: Michael O'Malley born 1859 – died 1946/ Lizzey O'Malley born 1860 – died 1917/ Mary Ann O'Malley born 1896 – died 1980/ RIP/ Erected by Michael O'Malley'

JOYCE, Catherine—see 7.
JOYCE, John—see 7.

8. *KIRBY, William*

Inscription: 'O Lord have mercy/ on the souls/ of/ William Kirby/ and his wife/ Mary Malley/ of Clare Island/ who died A.D. 1845/ Erected by their loving son/ John P. Kirby/ Citizen U.S. America/ T H Dennahy Glasnevin'

KIRBY, John B.—see 8.
MCCABE, James J.—see 4.
MCCABE, Cecilia—see 4.
MCCABE, Bernard—see 4.
MCCABE, Peter J.—see 4.
MCCABE, Michael J.—see 4.
MCHALE, Sarah—see 13.
MALLEY, Mary—see 8.

9. *MALLEY, Thady*

Notes: Noted by Westropp (1911, 37).
Inscription: 'O Lord have mercy on the/ soul of Thady Malley who de/ this life 28 Oct 1826 aged 84 year/ Erected by his sons John/ Anthony & Michael Malley/ Also the soul of John Malley/ who dep this life 12 May 1828/ aged 48 years'

MALLEY, John—see 9.
MALLEY, Anthony—see 9.
MALLEY, Michael—see 9.

10. *Oh(il)bin, Myles*

Notes: At the holy well (no. 151): see Pl. II. Noted by Westropp (1911, 28), who gives the date 1701. Perhaps the surname is Philbin.
Inscription: 'MIB.V/ IHS/ Peace and/ good / will/ (t?)o (all?) men/ 17(90?)/ Pray for Myles Oh(il)/ bin'

11. *O MAILLE*

Notes: Within the Abbey. An undated but elaborately decorated rectangular stone slab set in a moulded frame. For details on its iconography and dating, see Gillespie and Ó Comáin (this volume, pp. 41–44; Vigors 1893, 454; Westropp 1911, 30, pl. II).
Inscription: 'Terra mariq potens/ O Maille'

12. *OMAILLN*

Notes: Found in south-eastern corner of the graveyard in 1990 when digging the grave for John Martin O'Malley: now in National Museum of Ireland. An irregularly-shaped but parallel-sided sandstone slab (0.85m long, 0.32m wide, 0.07m deep) bearing a faint and roughly incised inscription on one face (Fig. 2).
Inscription: 'hy(c?) yace(t) (o?) ihe us oma(ill)n ridh ath (a?)s'

13. *O'MALLEY, Anthony*

Notes: Two associated memorials.
Inscription: 'Of your charity/ pray for the soul of/ Anthony O'Malley/ who died Dec 28th 1883/ aged 47 years and his wife/ Sarah nee McHale/ who died June 12th 1918/ aged 73 years/ also their daughter/ Mary Anne died June 23rd / 1913 aged 32 years/ RIP/ Erected by their loving son/ Austin O'Malley'
Inscription: 'Austin O'Malley/ 1875-1951/ Margaret O'Malley/ 1875-1957/ RIP'

14. *O'MALLEY, Anthony*

Notes: Two associated memorials.
Inscription: 'O Lord have mercy/ on the soul of/ Anthony Malley/ who died 16th Decr 1891/ aged 33 years/ RIP/ This stone was erected to his/ memory by the people of/ Clare Island as a token/ of esteem and affection/Monaghan Westport'
Inscription: 'In loving memory of/ Martin O'Malley/ died 8th June 1943 aged 83 years/ and his wife Sarah/ died 4th Feb 1951 aged 82 years/ RIP/ Erected by their family'

15. *O'MALLEY, Austin*

Notes: Within the Abbey in 1985: not located in December 2001.
Inscription: 'Pray for those who helped/ in the renovation of this church/ especially our emigrants in America and England/ Stations of the Cross donated in memory of/ Austin & Margaret O'Malley/ the Alter and Tabernacle/ in memory of Edward & Mary O'Malley The Mill'

16. *O'MALLEY, Austin*

Inscription: 'O Lord have mercy on/ the soul of Austin O'Malley/ who died August 8th 1874/ aged 45 years/ erected by his father/ William O'Malley'

O'MALLEY, Austin—see 13.

17. *O'MALLEY, Bridget*

Notes: Noted by Westropp (1911, 37) as being in the chancel of the Abbey: he gives the name as 'Bridget (Bratch)' and her age as 48. Within the Abbey in 1985: not located in December 2001.
Inscription: 'O Lord have/ mercy on the/ soul of Bridget O('Malley?)/ alias/ *[stone fractured, one and a half lines illegible]* ...(5?) 1802 aged 4.../ This

stone erec/ted by her hus/band Edmond O'Malley'

O'MALLEY, Edmond—see 17.
O'MALLEY, Edward—see 15.
18. O'MALLEY, John
Notes: Noted by Westropp (1911, 37) as being in the graveyard. Not located in 1985 or in December 2001. Westropp gives the following details:
Inscription: 'John O'Malley, died 1828, aged 48'

19. O'MALLEY, John
Notes: Noted by Westropp (1911, 37) who gives John O'Malley's age as 51. For the obscured line he gives 'by Peter Barrett'.
Inscription: 'Pray for the soul/ of John O'Malley/ who dep/ (2)1 June/ 18(4)2 aged 81 years/ also/ Walter Barrett/ who depd 2 May 1837/ (aged) 19 years/ [another line obscured by cement setting]'

O'MALLEY, Lizzey—see 7.
O'MALLEY, Margaret—see 13 and 15.
O'MALLEY, Martin—see 14.
O'MALLEY, Mary—see 15.
O'MALLEY, Mary Ann—see 7 and 13.
O'MALLEY, Nichl.—see 20.
O'MALLEY, Michael—see 7.
O'MALLEY, Sarah—see 14.
O'MALLEY, William—see 16.

20. O'MALLEY, Thos.
Notes: Noted by Westropp (1911, 37) as being in the chancel of the Abbey: he gives the brother's name as Michael. It is now recumbent in the floor of the church at the junction of the nave and chancel.
Inscription: 'The Lord have mer/cy on the soul of/ Thos O'Malley dece/ asd Janry 1st 1807 agd 34 yr/ This stone erectd by/ his brother Nichl O'Malley'

TOOLE, Charles—see 21.

21. TOOLE, Edward
Notes: Noted by Westropp (1911, 37) who gives Edmond rather than Edward.
Inscription: 'O Lord have mercy on/ the soul of Edward/ Toole who depd this/ life 22 July 1837 aged/ 21 years erected by his father Charles Toole'

Acknowledgements

Thanks are due to Bernie Winters of Kill for his ever generous hospitality and his knowledge of the ecclesiastical remains at Kill and to John Towler and Michael Savage for their invaluable assistance with the survey of the graveyard.

NOTES

1. The numbers accompanying each item provide the reader with a cross-reference to the detailed 'Catalogue of archaeological sites, monuments, and find spots on Clare Island', in P. Gosling, C. Manning and J. Waddell (eds), *New Survey of Clare Island. Volume 5: archaeology* (Dublin, Royal Irish Academy, forthcoming). The Abbey itself is no. 148.

REFERENCES

Cantwell, I. 2002 *Memorials of the dead: counties Galway & Mayo (western seaboard).* Irish Memorial Inscriptions, Vol. 1. CD-ROM. Dublin. Eneclann Ltd.

Vigors, P.D. 1893 Reports from counties: County Mayo. *Journal of the Association for the Preservation of the Memorials of the Dead, Ireland* **1**, no. 4, 452–7.

Westropp, T.J. 1911 Clare Island Survey: history and archaeology. *Proceedings of the Royal Irish Academy* **31** (1911–15), section 1, part 2, 1–78.

THE Ó MÁILLE MEMORIAL PLAQUE AND ITS HERALDIC ACHIEVEMENT

Fergus Gillespie and Micheál Ó Comáin

ABSTRACT

The plaque on the north wall of the chancel of the Abbey is described from a heraldic point of view and considered in relation to other O'Malley arms. The possible symbolism of the charges is discussed, and a dating for the plaque is suggested on the basis of comparable plaques in the west of Ireland.

The plaque and other O'Malley (Ó Máille) arms

The plaque (Pl. I) is built into the north wall of the chancel immediately west of the tomb niche. The arms are carved in relief on a single rectangular slab of limestone (915mm high x 630mm wide), which is set within a moulded frame formed of four long narrow pieces of limestone (see Manning, this volume, fig. 4, p. 15, for a cross-section of the moulding). The arms show: *a boar passant between three longbows charged with arrows and bent pointing in centre two in chief and one in sinister base and a quarter in dexter base charged with a skiff, the oars in action.*[1] The crest is: *a horse in full gallop* and the motto: *TERRA MARIQUE POTENS.*

This heraldic achievement resembles very closely the arms of O'Malley of Rosehill,[2] County Mayo, which are: *argent, a boar passant gules, bristled or, between three longbows charged with arrows and bent pointing in centre one in chief and a skiff with oars sable between the two longbows in base.* This branch of the family is descended from Brian mac Cormaic Ó Máille of Murrisk, *fl.* late sixteenth century (Burke and Burke 1844). However, another version of these arms has an ancient galley rather than a skiff in base.[3]

Other O'Malley arms have only the boar: *or a boar passant gules,* with the crest: *a horse rampant argent.* These arms are those of O'Malley of Snugburough,[4] who was descended from the Belclare branch. Sir Owen O'Malley (born 1887), a distinguished member of the British Foreign Service, was of this branch. According to Burke's General Armory (Burke and Burke 1844) these were the original arms of Ó Máille but with a ship of three masts, sails set proper as a crest.

The arms on the slab are curious. The addition of the quarter in the dexter (heraldic right, viewer's left) base appears to be do-it-yourself Gaelic heraldry and very much out of line with other Irish heraldic practice. It is also possible that the craftsman did not understand the heraldic illustration he was working from, and the arms on the slab could be a badly executed version of the arms of O'Malley of Rosehill (see blazon above).

The charges and their symbolism

The boar is the principal charge of the arms and is clearly being hunted. Beasts of prey are common enough as charges in heraldry, but here the boar may reflect one of the many sovereignty myths found in early Irish texts. These often tell of a hunt of a wild animal, such as a boar or stag, by candidates for kingship. The hunt moved round the borders of a territory, and the young man who captured the prey became its new ruler. This royal hunt is well documented in early Irish literature.

Pl. I   The Ó Máille plaque in the north wall of the chancel (DOEHLG).

After the death of Echu Mugmedón, king of Tara, his sons, one of whom is Niall Noígiallach, are sent on a hunt by the smith, Sithchenn, who had just forged new weapons for them for the task. Shortly after eating the captured prey, Niall is declared king by a female figure of Sovereignty who he encounters at a well, and with whom he has had sexual intercourse despite her revolting aspect (Stokes 1903). In a poem version of the same tale the brothers make a circuit of Ireland in pursuit of a boar, which they slay and eat, and after this, as in the prose version, the figure of Sovereignty promises Niall the kingship of Tara. A similar story is told of Lugaid Laígde of the Érainn and his brothers. Here a druid prophesies to Daire Droimthech that a son of his, whose name would be Lugaid, would become king of Ireland after killing a fawn that would come into the assembly. The king as a result calls all his sons Lugaid. When the fawn eventually appears the brothers pursue it. At Binn Étair Lugaid Laígde, separated from the hunt by a mist, finds and kills the fawn, after which he too sleeps with a hag (again a Sovereignty figure) who promises him the kingship of Ireland (Stokes 1897). As recently as four years ago Fergus Gillespie heard a similar account regarding the hunt of a stag from an old man in the Inishowen peninsula in County Donegal that related to the lordship of O'Doherty, whose arms are *argent a stag springing gules on a chief vert three mullets argent*. The stag is also found in the arms of other Gaelic families, the best known being those of Mac Cárthaig, a branch of the Eoganachta. Because the hunted animal and the hag/Sovereignty figure are mutually exclusive in these sovereignty tales, it is generally accepted that the hunted prey and the lady are one and the same element.

The theme of the hunt of an animal as a prerequisite to accession to kingship is not only found in Irish literary sources, but also in Welsh and Breton literature (see Bromwich 1961 for discussion; for a brief discussion of pre-heraldic themes in Gaelic heraldry see Gillespie 2000, 12–13).

The ship may represent the maritime power of the family, which was noted for its raids by sea on neighbouring territories: for instance in 1513 an Ó Máille fleet attacked Killybegs in the territory of Mac Suibhne Boghaineach in south Donegal. This maritime prowess is also reflected in their motto *Terra marique potens*, 'Powerful on land and at sea'. The horse for crest is intriguing, and one can only guess at its symbolism. Gaelic arms often allude to remote ancestors, and there is a good case to be made for its representing the remote ancestor of the Ó Máille, Echu Mugmedón, father of Niall Noígiallach and king of Tara: Echu, a diminutive of Echaid, means 'a horseman'.

Discussion and dating

The ship is a powerful symbol of the O'Malley family's maritime tradition, but its inclusion on the shield on an imposed quarter (or canton?) is heraldically eccentric to the extent that it almost appears to be an afterthought or correction by the carver. A canton, which is smaller than a quarter, is, in Irish armory, always found at the top of the shield, and while a single quarter, in theory, can be placed in base, this is highly unusual. Cantons in base are not unknown in Continental usage, but it would be to risk over-interpretation to posit a Continental influence in this case.[5] It is certain, however, that this armorial achievement did not originate in the Office of the Ulster King of Arms, so it is difficult to say with any certainty to whom it appertained.

Memorial plaques with coats of arms carved on them are common in the seventeenth century in the west of Ireland but have not been studied as a group. Micheál Ó Comáin has observed one example strikingly similar to the Clare Island plaque—the Burke coat of arms carved in stone in Claregalway friary, Co. Galway (Cochrane 1901, 328), which has been dated to the 1650s. Plaques in Rosserilly Friary, Co. Galway (one a MacDonnell memorial dated 1646 and the other a similarly dated Burke memorial) are also comparable.[6] As the mantling and tassels are similarly arranged we can only conclude that the Clare Island plaque was executed either by a sculptor from the same school or at least a sculptor who had seen these carvings or a common model. The similarities are most obvious in the treatment of the helmet: the mantling, which is not attached to the wreath on top of the helmet, comes out laterally as strands before falling as plaited cords, which end in tassels. This is true also of the treatment of the arms of Ó Máille impaling Browne carved on a stone plaque

(Pl. II) in Murrisk Abbey, on the southern shores of Clew Bay, which was probably commissioned to commemorate the marriage of Captain Eoghan Mór Ó Máille to Martha Browne of the Neale in 1675 (Corlett 2001, 82). It bears the date 1719, but this may be a later addition. This plaque is either by the same hand as the Clare Island plaque or from a common pattern. If we take it that the sculptor was following a definite tradition in stone carving there is nothing to prevent our dating the slab to the latter half of the seventeenth century or the beginning of the eighteenth century. However, a date closer to that of the marriage (1675) is more likely.

Pl. II The plaque of Ó Máille and Browne at Murrisk Abbey (Chris Corlett).

NOTES

1. Colours: the tincture or refers to gold, argent refers to white, gules to red, sable to black, and proper to natural colouring. Positions: in chief is at the top part of the shield and base at the bottom. Dexter is the heraldic right or viewer's left, while sinister denotes the heraldic left or viewer's right. Attitude of charge: passant when referring to an animal means that it is walking. Divisions of shield: a quarter is one of four or more roughly equal divisions of a shield; a canton is a square smaller than a quarter. An impaled shield is one that displays two coats of arms separated by a vertical line.

2. The ruin of Rosehill House is in the townland of Money, parish of Kilmeena, near the shore of Clew Bay north of Westport (OS six-inch sheet 76). For a short account of this O'Malley family see Mulloy 1988, 38.

3. Genealogical Office MS 105, Grants and Confirmations C, fol. 26, 1804.

4. The townland of Snugborough (OS six-inch sheet 78) is situated a short distance west of Castlebar (Mulloy 1988, 80).

5. We are grateful to Conleth Manning for his suggestion that as Sir Charles O'Malley spent most of his life in continental Europe he may have introduced this variant. However, the most comprehensive European armorial (J.-B. Rietstap 1967) does not include these arms of O'Malley but instead includes the more usual form of the arms as found in Murrisk Abbey, where it is not the O'Malley arms but the impaled coat of Browne that is found incorrectly exemplified.

6. We are grateful to Professor Roger Stalley for bringing the plaques at Rosserilly to our attention.

REFERENCES

Bromwich, R. 1961 Celtic dynastic themes and the Breton lays. *Études Celtiques* **9**, 439–74.

Burke, J. and Burke J.B. 1844 *Encyclopaedia of heraldry, or general armory of England, Scotland, and Ireland*. 3rd edn. London. Henry G. Bohn.

Cochrane, R. 1901 Excursion to County Galway. *Journal of the Royal Society of Antiquaries of Ireland* **31**, 305–40.

Corlett, C. 2001 *Antiquities of West Mayo*. Bray. Wordwell.

Gillespie, F. 2000 Heraldry in Ireland: an introduction. *The Double Tressure: Journal of the Heraldic Society of Scotland* **23**, 7–25.

Mulloy, S. 1988 *O'Malley people and places*. Whitegate and Westport. Ballinakella Press and Carrowbawn Press.

Rietstap, J.-B. 1967 *Illustrations to the Armorial Général by V. and H.V. Rolland*. London. Heraldry Today.

Stokes, W. (ed.) 1897 *Cóir Anmann. Irische Texte* **3**, 2 Heft, 317–23.

Stokes, W. (ed.) 1903 *Echtra Mac Echach Muigmedóin. Revue Celtique* **23**, 190–207.

INTRODUCTION TO THE WALL PAINTINGS

Karena Morton and Christoph Oldenbourg

Fieldwork and research are revealing more and more traces of wall paintings on medieval buildings in Ireland, though the traces are often small and fragmentary. Our knowledge of these wall paintings has greatly increased as a result of a number of conservation projects commissioned by the National Monuments Service at monuments such as Cormac's Chapel at Cashel, Jerpoint Abbey, Abbeyknockmoy, Ardamullivan Castle and, of course, the Abbey on Clare Island. It can confidently be said that the Clare Island paintings are unique for several reasons. The Abbey has two painting schemes—a rare occurrence in Ireland. Comparatively speaking, these painting schemes are relatively complete: many of the extant Irish wall paintings are known only from a few fragmentary remains. Moreover, although the subject matter of the later scheme of wall paintings at the Abbey falls within the general European tradition, the individual style or character of the images, and probably therefore their significance, appears to be unique.

The Abbey on Clare Island is the most westerly presence of the Cistercians in Europe. It was a daughter cell of Abbeyknockmoy in Co. Galway. The earliest mention of the church being painted is in a book on the Cistercian Abbeys of the world published in 1640, but no details of the paintings themselves were given in that account (see Manning, this volume, pp 7–10). The surviving paintings are confined to the chancel area of the Abbey. T.J. Westropp first documented them as part of the original Clare Island Survey in 1909–11 (1911, 29–37). The work Westropp carried out on the wall paintings was comprehensive and as accurate as circumstances allowed. Tracings were made by the only means available, 'from a platform of barrels and planks—a weary, and painful task' (1911, 29). The condition of the Abbey and the surviving visible wall paintings were described. Westropp, who clearly was working under difficult conditions, correctly recognised two painting phases, although he was

confused in ascribing certain parts of the images to one phase or the other (1911, 32).

Around the time of the original Clare Island Survey a cement floor was inserted over the top of the vault and a few other small repairs were made. In the eighty years after the completion of the survey the paintings continued to deteriorate. The re-roofing of the chancel in the winter of 1952 did not greatly improve the situation. At that time the ground around the Abbey was excavated to original floor levels. Paradoxically, the paintings themselves appeared to deteriorate more rapidly after these interventions. This rapid visual deterioration was due to the change in microbiological habitat. Dense, green/black growth established a foothold on and within the fabric of the plaster and paintings—more or less concealing them from view.

In 1990 the National Monuments Service initiated a major conservation programme. Between 1991 and 1999 this project was carried out and directed by Madeleine Katkov and Christoph Oldenbourg, with the assistance of many conservators, including the co-author. Now far more of both painted schemes is visible than was previously recorded.

The conservation programme afforded the opportunity, over a number of years, for detailed scrutiny of every aspect of the paintings by a good number of pairs of eyes at different stages in the treatment of the paintings. The scaffolding platform was the venue for many detailed and animated discussions, where hypotheses could be checked and double-checked right in front of the painting. Such a privileged position was not an option for Westropp. It should be noted that he had only five days to work on every aspect of the Abbey. In the circumstances his achievement was immense. The sheer size of this task, the primitive technology available to Westropp (for example the lack of decent lighting), together with the condition of the paintings and the rudimentary state of knowledge about wall paintings in general at that time, made for a complicated undertaking. It is hardly surprising that Westropp did not recognise certain important facts about the painted decoration of the Abbey or that there are some errors of judgement, omissions or misinterpretations.

Despite these limitations, Westropp's extensive documentation, which includes illustrations and written and photographic records, is an invaluable archive. It was and continues to be part of the primary record. It served as a benchmark in the initial reading of the images during the recent wall painting conservation programme and is heavily drawn upon in all of the papers on the wall paintings in this volume.

The main work of the conservation programme is now completed, and this has roughly coincided with the publication of the New Survey of Clare Island. It is hoped that the information gleaned through the programme and presented in these studies will make a valuable contribution to the survey. The insights gained should inform future debates on the many and varied aspects of the wall paintings in the Abbey.

The primary aim of the following four separate but linked studies is to provide an detailed up-to-date account of the nature and extent of the wall paintings.

There are two distinct painting phases. The earliest painting scheme, perhaps originally created in a chancel with a flat wooden ceiling, is found today on the north and south walls and extends as high as the springing of the present vault. The vault was added in a secondary construction phase and consequently only has paintings of the second scheme, which, as well as occurring on the vault, survive to a lesser extent on the east wall and in an even more fragmentary state on the south and north walls. The paintings will be consistently described: the earlier paintings are referred to as Phase One, while the later scheme of painting is referred to as Phase Two.

In the first paper, one of the conservation directors (Christoph Oldenbourg) gives an overview of the condition of the wall paintings before conservation, an account of the original materials and techniques and a discussion of the decay mechanisms, together with a summary of the conservation treatments. The second contribution is a catalogue of the wall paintings with detailed records of the images as they survive today. A comparison between what was visible or recorded in 1909–11 and what is currently visible is included. The catalogue is intended to augment and update the work carried out by Westropp some ninety years ago. The third contribution considers the iconography and dating of the wall paintings. The materials and techniques employed for the wall paintings at Clare Island are commonly found throughout Europe. The iconography of the Phase Two paintings, however, is unusual in the mode of representation rather than its subject matter or the messages conveyed. As yet no other wall paintings have been found in either an Irish or European context with a similar scheme. In this respect the Clare Island wall paintings are, so far, unique. Certain theories and symbolisms are suggested for consideration. An overall theme for the subject matter is proposed. This proposal is based firstly, on the isolation and study of each individual component and secondly, on a combined assessment of all of these elements to establish a common denominator and therefore, perhaps, an overall theme. The fourth study is on the musical instruments depicted in the paintings, two of which were uncovered by this conservation work and are published for the first time here.

A detailed account of the architecture of the Abbey is not included here, as this is addressed in another contribution (Manning, this volume, pp 7–27). However, given the stratigraphic relationship between the wall paintings and the supporting fabric, certain information on the building phases and a possible dating were gleaned during the course of conservation. A relative chronology for the building phases and alterations, as suggested by the wall paintings themselves, is proposed.

REFERENCES

Westropp, T.J. 1911 Clare Island Survey: history and archaeology. *Proceedings of the Royal Irish Academy* **31** (1911–15), section 1, part 2, 1–78.

CONSERVATION OF THE WALL PAINTINGS

Christoph Oldenbourg

ABSTRACT

The conservation work was carried out on the wall paintings between 1991 and 2001. An international team of scientists and conservators was brought together, managed and co-ordinated. Two relatively complete cycles of paintings were recovered from an extreme state of decay and virtual collapse. The condition of the paintings, the techniques with which they were created, the decay mechanisms affecting them and the conservation processes used are described, as is the presentation of the paintings.

Introduction

The recent conservation of the Abbey was started in 1991. It is not clear what, if anything, was done in the nineteenth century for the upkeep of the Abbey. Several interventions to protect the building were carried out in the twentieth century. Not all of these were successful. In this article the various factors contributing to the deterioration of the Abbey will be discussed. A summary of the decay mechanisms and the conservation treatments is also included. A full conservation report will be lodged with the clients—The National Monuments Service—and will be available to those requiring a more detailed technical account.

Overview of condition before conservation
Condition of the Abbey and paintings in 1911

The recorded building interventions as well as T.J. Westropp's comments on the condition of the paintings at the time of the first survey (in 1909–11) allow some insight into the history of the deterioration of the wall paintings. From Westropp's account we can glean information about the effects of introducing cement repairs into a historic building and we can deduce information about the process of calcification on

the surface and the degree of biological growth at that time.

The Abbey was taken into state care in the 1880s, when the Office of Public Works made some tentative attempts to arrest the ongoing decay, but specific records of these works were not kept. It is known that a concrete floor was inserted over the vault at some time in the years between 1908 and 1910. A hole was broken into the south wall to allow the water to drain away from that floor. This did not prevent water seeping through into the vault below and onto the paintings. Westropp remarked that 'The Board of Public Works has done everything to preserve the ceiling, but in vain; the damage had too long set up from water soaking through the floor above for anything now to be effective' (Westropp 1911, 33). Westropp also reported on a comment made to him by the Rev. E.A. Lavelle, curate on the island, who in turn had been told that the paintings were in good preservation in 1862 when the new church was built (Westropp 1911, 33). If this report can be taken as accurate, it suggests that deterioration of the Abbey and wall paintings accelerated from the mid-1860s.

Westropp noted the fading out of a colour in

the Phase Two paintings in the vault. It is most likely that the 'blue' recorded by him is actually a layer of calcification beginning to build up on the surface, such as can be seen today over the figure of St Michael. Westropp also observed a 'chocolate brown', which was more than likely a biological staining. In his descriptions of the vault paintings Westropp differentiated between that which had been lost— 'bare of plaster'—and that which was concealed—'destroyed…by green growth' (Westropp 1911, 35). This distinction is important, since many of the images 'destroyed…by green growth' were recovered during the recent conservation programme.

Westropp also noted that hardly a trace remained of any design on the side walls: 'The older painting, however, shows nearly everywhere if a flake has fallen from the later plaster; but no pattern is discernible…' (Westropp 1911, 35–6). Again, it is clear from this statement that sufficient of the Phase One paintings survived despite being masked by micro-biological and other decay products, by the raised interior floor level and by the overlying later paintings.

Westropp had felt that the deterioration of the paintings was so severe that there was unlikely to be any measure that could be undertaken to save them from their ultimate demise (Westropp 1911, 33). Indeed, it is probable that this belief accounts for the extremely detailed records of the wall paintings that Westropp made. He must have worked on the premise that the paintings would soon be lost completely and the recording of them was of the utmost importance. In this light, one might have assumed that the paintings would have long since perished. Fortunately, although deterioration progressed steadily, surprisingly little of the wall paintings has been actually lost in the intervening years.

The condition of the Abbey and paintings between 1911 and 1991

By 1952 the Abbey had been a ruin for several centuries. In the winter of that year the Office of Public Works re-roofed the chancel, removed the concrete floor over the vault and excavated the area in and around the church down to the original floor levels. On the interior, loose plaster edges were secured with mortar, perhaps also in 1952. Thankfully, this was mainly done with lime

plaster that contained only small amounts of cement, if any at all. The repair mortar is grey (because dark grey limestone sand was employed in its make up) and might therefore be interpreted incorrectly as cement.

In 1991, the year the most recent conservation began, the walls were saturated with water and covered with thick layers of microbiological growth, which together with calcite deposits obscured most of the paintings on the vault and nearly all of the Phase One paintings. The plaster detachment had progressed dramatically, and large sections were in imminent danger of falling.

Original materials and techniques

The altar, the window frames and the chancel arch are made from local serpentinite. Only the O'Malley armorial plaque and the traceried canopy over the tomb niche are of limestone, which was imported from the mainland. The masonry was bonded with a very simple lime mortar, using local sand, albeit not very well graded. With this the stones were laid, the pointing effected and the render and plaster applied. The paintings were also executed using lime as the paint medium. The source of the lime has not yet been identified, but it was most likely derived from local limestone.

Building techniques

The rubble walls are made from local stone. As a geological fault runs through the island, a variety of stones can be found. These were roughly dressed but were more finished around quoins and windows, and were well laid and set in lime mortar. The pointing mortar, quite grey in colour, was cursorily levelled, apparently with a brush. The walls were plastered with a finer mortar, of a slightly lighter grey colour than the pointing. This was worked with a brush and spread very thinly in places. The walls were finished with a limewash to receive the paintings (Pl. I).

The insertion of a pier in the north-east corner was necessary as a first step in the construction of the vault. A relieving arch surmounted this pier. A barrel vault was then constructed over wicker centering. Frequently the wicker centering used in the construction of a vault is simply left in place and plastered over. At Clare Island only a few fragments of the wattle used in the centering were found, and it might be postulated that for

Pl. I   Phase One, detail of the stag. The rubble wall fabric and the various plaster layers of the Phase One paintings are clearly visible, as are details of the painted and incised design.

technical reasons the centering was removed once it had served its purpose (Pl. II). This suggestion would help to explain the very good survival of the vault plaster despite the extreme conditions. Wicker centering was noted elsewhere in the Abbey: the south-wall window head retained the impression of wicker, which was exposed where plaster had fallen away. This and similar depressions noted elsewhere in the church were covered by new, soft lime mortar during recent repairs to the painted plaster.

A coat of levelling plaster was applied to the support, which was light in colour and of medium-coarse consistency. Its thickness varies considerably but it is 40mm thick on average. A fine, white, lime-rich finishing plaster some 4mm deep was then applied to receive the paintings. With the exception of the paint on the stonework of the tomb and canopy, these Phase Two plaster layers covered all of the Phase One paintings. It should be noted that the plaster on the back wall within the tomb recess most likely belongs to Phase Two on account of the close resemblance (in terms of colour, texture and aggregate) to the plaster of these later paintings. Analysis to verify this has still to be carried out.

Painting techniques

Both painting phases are lime paintings rather than frescoes.

Phase One paintings

The design was mapped out by incising the plaster, which was still partly damp when this process was carried out. A sharp instrument such as the end of a brush was used for this. The pigments were painted onto fresh limewash or applied with a medium onto dry plaster, i.e. *a secco*, not directly onto fresh wet plaster.

Pl. II   In this view the impressions left from the wicker centring and the various plaster layers that form part of the Phase Two paintings in the vault area can be seen.

The surviving pigments of this painting phase are carbon black, an exceptionally good-quality red ochre and lime white. No traces of wasted pigments were found from this first painting phase. It is of interest to note that the face of the horseman on the south wall is black. This black seems to be the same black as the armour rather than a discoloured pigment such as converted lead white. Some design elements and corrections, such those made to the coat of mail of the horseman on the south wall, were scratched into the dry painted plaster afterwards. 'Pentimento' is the term used for a change in the design during the process of laying-out or painting a picture. In the Phase One paintings at Clare Island Abbey, the stag on the north wall was initially sketched as halting or stationary, with at least one front leg firmly placed on the ground, whereas in the finished version the stag is painted in full flight with both front legs stretched out.

Phase Two paintings

As in the earlier scheme the paintings of Phase Two were first sketched by incising the plaster. As a first step the black false ribs were set out. A flexible batten was probably used as a straight edge, and again the end of a brush or a spatula was used to deeply incise the lines. Further incised lines divide the ribs into masonry blocks with a lozenge at the apex where the diagonals cross. Their blue-black hue and slightly greasy texture indicate lamp black (soot), rather than charcoal black, which has a browner hue and is less greasy. For the design of the ceiling bosses a compass was used. The figurative scenes too are loosely based on incised drawings, executed quite rapidly with a very fine point in a moderately smooth plaster surface that seems not to have been quite dry. Some of the designs are only visible by virtue of the incised lines. Analysis to determine whether they were originally coloured or whether the colour has since deteriorated has still to be carried out. Images that were originally painted with less strong colours were more easily obscured by calcite and mineral deposits.

Again there are a few pentimenti to be found on the Phase Two wall paintings. Perhaps the most obvious are to be found along the painted ribs, where initial lines incised for the ribs have subsequently been altered to accommodate the scenes in the interspaces (Pl. III). Sometimes,

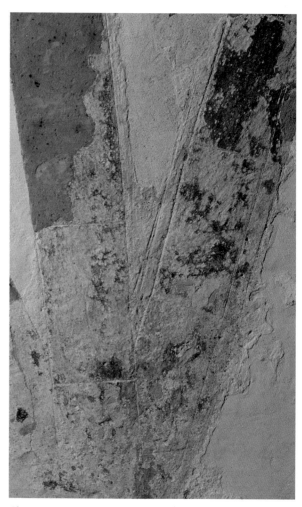

Pl. III   The corrections or pentimenti in the placing of the ribs are clearly visible here.

however, the positions of the ribs were not changed and the figures overlap the ribs.

The colours used in the Phase Two paintings were also applied *a secco*. The pigments are of the standard medieval range, and what survives is of exceptionally good quality. It seems significant that the highest quality pigments were obtained by, and affordable to, the Clare Island painters. The colours are predominantly a strong bright red and a clear yellow. In view of the water penetration and migration of lime that these paintings have suffered over the centuries, their survival can only be accounted for by their exceptional quality.

Yellow and red ochre is used for the figurative designs, together with lime white. No black is employed in the design of the images themselves. Although Westropp referred to a chocolate brown colour, the only chocolate brown to be seen at the start of conservation was a stain from bacterial growth.

In one or two instances, especially in the floral design of the pier on the east wall, UV fluorescence showed up a further distinction. This is very faint indeed but may well be the residue of a medium (possibly water-soluble) that was used in the *a secco* application of a colour that is now long gone. There is also a great deal of faint grey and grey–brown; Boss D, on the ceiling, seems to provide evidence that there is more than one faded colour. Indeed, Westropp mentioned a faded-out colour, which he suspected was originally purple or light blue; for brevity he called this blue. It is difficult to gauge precisely what quality of colour he referred to in these instances, but it is possible that he mistook the first developing grey efflorescence as a medieval colour. Certainly, the most frequent occurrence of the colour blue as described by Westropp is at the west end of the vault, in exactly the same position as the large patch of grey efflorescence that masks the images today (most extensively around St Michael).

In other instances Westropp employed the term blue for a colour that today appears beige, such as the griffin in Section F, on the south side; this would most probably have been a mix of red ochre with white lime.

Decay mechanisms

The single most destructive substance to historic fabrics is water (Raschle 1996). It can enter a building in many ways: as precipitation from above and as rising damp from below. It can be absorbed into the fabric as run-off from the roof or driving rain. It can be actively drawn into the fabric through hairline cracks in pointing that is too hard or is failing, and it can be carried into the building by condensation. Once within the fabric, water can dissolve the building materials. Under wet conditions it can leach out and carry damaging substances further into the fabric in their dissolved liquid form; in drying-out phases it will draw these substances onto the surface again. As ice, it will split or blow substrates, as will other materials that are transported by water through the building fabric. In addition, water is the necessary prerequisite for any form of microbiology and for higher forms of plant life, which damage a building in their numerous individual ways. In the case of Clare Island, water entering the building carried effluents from the

soil that had accumulated everywhere in and around the building and leached sulphuric substances out of the cement floor or roof. These, together with salt contamination from the nearby sea, produced a potent cocktail of destructive substances. Soft rainwater leached acids out of the soil that had accumulated on top of the vault. Soil acids slowly dissolved and eroded the building limes, thereby weakening the substrate. The dissolved lime then percolated to the plaster surface. In contact with air, the lime reset again, forming a hard opaque crust that obscured the surface, in this case the paintings. It created an unnaturally hard surface over a now weakened substrate.

A further source of lime is cement. As lime is a reaction product of the hydration of calcium silicates (the main constituent of cement), it will be present wherever cement is employed. Older cements will yield up to 3% free lime, and even more-modern cements will still yield up to 1% free lime. This situation is compounded by the fact that hydrated lime is frequently added to cement mortars to make them more workable. This additive has no effect on the setting reaction of the cement at all. It will therefore remain soluble and can potentially be mobilised by water. Coke-burnt cements, such as the one used at Clare Island in 1908, will always contain quantities of sulphur. This highly reactive substance will convert any free lime into gypsum. Gypsum crystals are bigger in volume than lime and will therefore blow the plaster (Pl. IV) Furthermore, the sulphuric effluents of cement will also react with many other substances, creating numerous new minerals such as natrite ($NaCO_3 \cdot 10H_2O$) and mirabilite ($NaSO_4 \cdot 10H_2O$) (Pl. V). Mirabilite characteristically occurs in the presence of

Pl. IV Large gypsum spheres in the head of the dragon in Section E on the north side of the vault.

gypsum, which was created by the cement in the first place (Bläuer Böhm 1994) (Pl. V).

Substances that dissolve and recrystallise when undergoing a wet and dry cycle are generically termed salts. These salts could come from the sea, as in the case of Clare Island, as well as from decomposition products of soil etc. When dissolved in water, salts are carried into porous building materials. As the building undergoes a drying-out cycle, the salts will crystallise, increasing in volume, with plaster disruption as an inevitable consequence. The failing pointing, as well as the general use of the Abbey as a donkey shelter and playground, all played their part in the deterioration of the wall paintings. The ubiquitous moisture had saturated the fabric: together with the then (pre-1952) raised floor levels (1.2 metres on average) this supported a rampant microbiological growth, which further weakened the precious painted plaster on the interior (Pl. VI). The insertion of the cement floor (later removed in 1952) was certainly a very mixed blessing. Although it must have kept out the rain for some time, it would have increased condensation on the interior (painted) surfaces considerably, as cement is mildly hygroscopic. This in turn would have encouraged microbiological growth.

Microorganisms not only obscure the surface themselves but, in some cases, produce minerals and calcification, further concealing the surface (Pl. VII) (Petersen 1992). In the course of their life cycle the microorganisms become damaging and eventually provide nutrients for further growth. Several organisms first attach themselves to the surface by etching it with acidic excretions, thereby eroding the surface. Fungi and bacteria penetrate the pore structure of the substrate,

Pl. V   Crystalline formation of mirabilite in the north-west corner of the vault.

Pl. VI   Microbiological growth on the south side of the chancel before conservation.

Pl.VII   Detail of microbiological calcification.

thereby weakening it (Raschle 1996). They produce layers on the surface, which retain moisture and the damaging substances carried with the water. They act, therefore, not unlike a poultice or compress.

Most of these processes were active a long time before the first survey and have continued ever since, up to the 1990s. It is no surprise, therefore, that the painting looks worn and faded in places. More seriously, this has meant that the interior plaster had started to detach in its entirety. Large sections had fallen away already by 1909. In comparing Westropp's descriptions and watercolours with our own observation, we became aware of several losses that had occurred in the interim. What had survived to the 1990s was increasingly eroded and obscured to a large extent by microbiology. Given these circumstances, it is remarkable that so much of the wall paintings did survive.

Conservation processes

Preliminaries

Prior to conservation a detailed archaeological, technological and aetiological examination of wall paintings is normally undertaken. However, after initial examination of the Clare Island paintings in 1990, the necessity for immediate emergency repairs to prevent further losses was instantly obvious. As a result, only the most essential examinations were carried out and all further thorough study of the object was deferred until the wall paintings were adequately secured.

Following discussion with the site conservation architect and input from other disciplines, an integrated programme of conservation based on the principle of treating the building as a single entity rather than a container for the wall paintings was developed. The first-aid programme therefore was designed initially to alleviate the most extreme and obvious detrimental processes and building failures, i.e. to reduce water ingress, improve the water disposal and secure the most vulnerable elements of the object, in this case the painted plaster. In the context of a multidisciplinary approach we made sure that we backed up all our findings with scientific advice (see Petersen 1992; Bläuer Böhm 1994; Raschle 1996). Interventions were selected that would allow for gradual changes in the environmental conditions and could be monitored (and more easily controlled) while observing their success or otherwise. Too rapid a variation in conditions, it was feared, might have caused the object to undergo too sudden a change and might possibly result in damage.

The materials and techniques selected for use in these initial processes were chosen to be compatible with the original materials as far as possible. The newly introduced materials had to be of the same physical strength or preferably slightly weaker than the original, so as not to create further pressure on the already stressed original.

Every effort was made to avoid introducing new materials into an already extremely complex situation. Traditional building materials were used for the most part, including well-washed sand and high quality lime, in some instances with the addition of trass (a volcanic hydraulic material). The trass used at Clare Island was tested for its purity. The addition of trass enables lime to react in a hydraulic way. Hydraulicity describes the property of a mineral binder that can set when air is excluded. This property was essential when injecting the cavities behind the detaching plaster. Trass is the equivalent to pozzolana, which the Romans used to produce their famous cements. These additives do not have the drawbacks of modern cement.

Trass was also used to improve the resilience of the thin lime render or shelter coat that was applied to the exterior of the chancel. This proved to be the single most effective intervention in keeping the rain out of the building. It also infinitely improved the internal climate. The one modern material employed in the conservation of these paintings was a silicic acid dispersion, which sets to silicon dioxide, which is the main constituent of pure sand. In certain situations it is easier and safer to use than lime, for example to re-attach paint flakes or to consolidate friable plaster.[1]

Treatment of microbiology

Because of our resolution not to introduce new materials into the wall plaster, the microbiology was treated by irradiation with germicidal UV light rather than with biocides. This technique was pioneered, tested and evaluated by the author for the first time in the treatment of mural paintings at Abbeyknockmoy, Co. Galway. Germicidal UV light provides a method of killing most of the microbiology in a non-invasive way. It interferes with the cell structure of the organisms and/or with their genetic material. Under damp conditions it produces certain amounts of ozone, which also has a sterilising action. To be effective, the area to be treated is irradiated continuously for one hundred hours, by which time the organisms will have bleached and are unable to recover, dying completely within approximately ten days. The microbiology can then be removed in a conventional way as with other uncovering processes. The University of Oldenburg, Germany, under the umbrella of a Eurocare project, provided scientific support by analysing microbiological samples, establishing the effectiveness of the treatment methods (Pl. VIII).

For the first time in wall painting conservation UV fluorescence and infrared photography were used as a non-invasive way of monitoring the

Pl.VIII   UV lights set up for irradiation of the vault.

success of the irradiation treatment, as described above (Pl. IX). This photography was undertaken at regular intervals throughout the conservation programme, and the results of both methods were extremely informative.

Plaster consolidation

The most pressing conservation process was the consolidation of the detaching and delaminating wall plaster. The hollow areas were supported with fresco presses (Pls X and XI). Small access holes (generally 2mm diameter) were drilled

through the plaster in areas where the surface was already damaged. The voids between the layers of plaster were thoroughly flushed out and cleaned and subsequently injected with an adhesive grout. As indicated above, this grout was made up of lime putty and trass. Depending on the size of the void, inert fillers such as marble flour or very fine sand were added. As there are three material layers on the vault, delamination could occur at any or all of the interfaces. To borrow a phrase from Westropp, this process was 'a weary, painful task' for those involved, and over the years not

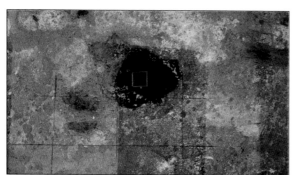

Pl. IX   UV fluorescence photograph of test irradiation. The red fluorescence indicates healthy green growth. This section had been covered prior to irradiation and therefore was not exposed to the UV light. The light grey sections indicate areas successfully treated. Here the microbiological growth has been killed.

Pl. X   Large areas of detached plaster were supported by fresco presses while the remedial mortar edging was removed. The presses were kept in place until the grouting mortar had set.

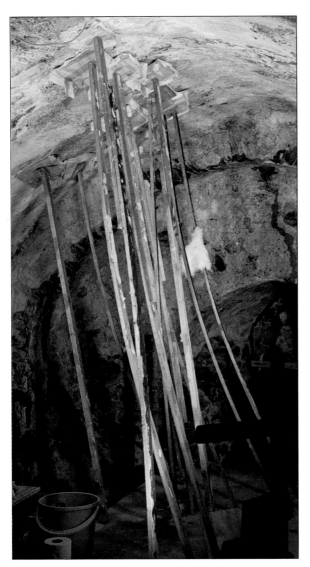

Pl. XII   The south side of the chancel (the same area as in Pl. VI) after conservation.

Pl. XIII   Removal of the microbiology with a small brush and conservation sponge.

Pl. XI   A forest of fresco presses was often necessary to support extensive areas during grouting.

less than seven full working months were spent on plaster consolidation and grouting alone.

The large areas where the plaster (and with it the paintings) had fallen away altogether were filled with a soft porous lime plaster. This was necessary in order to support the remaining original plaster, and by virtue of being more porous than the original the new plaster offered an easier route than the original for pollutants in solution. As a result the repair plaster will attract damaging substances in preference to the original, thereby protecting it. The alkalinity of the new mortar acts like a disinfectant, inhibiting biological growth, and being designed to be sacrificial, it can be removed and replaced should this become necessary in the future. Its colour and texture were very carefully chosen, and this helps enormously in the reading of the original images.

Surface cleaning

A great deal of skill, patience and perseverance was required for the cleaning process (Pl. XII). Contrary to the impression created by films such as *A Month in the Country*, conservation treatments take much time to achieve results, and the glamorous image sometimes associated with this profession is far from the reality.

At Clare Island Abbey the cleaning entailed the removal of the irradiated microbiology with small stiff brushes, sponges and alcohol/water solutions (Pl. XIII). Decayed support plaster that had lost all its cohesion, as well as lime encrustations, had to be removed in their entirety without endangering the precious paintings. The paintings concealed by small amounts of overlying limewash (in Sections I and J on the north side of the vault) were also uncovered.

Pl. XIV   The use of a mini-grinder during the uncovering.

The uncovering was carried out mechanically. Every conceivable dental tool came into use (Pls XIV, XV), since all chemical options proved ineffective, unreliable, too slow, or hazardous to the health of the conservators, or had all of these disadvantages.

Most areas were worked over several times before the task could be considered complete, as the original surfaces and paint layers had to be secured and fixed and re-fixed at several stages as cleaning progressed. In the summer of 2000 the stabilisation process was concluded.

However, at the time of writing this article, the cleaning and uncovering process is still incomplete in a few areas. In these areas it was not possible to use conventional techniques without endangering the paintings. This led to the testing of a laser in the late summer of 2000. For many of the problems encountered the results were very promising and the legibility of the images was greatly improved.

Presentation of the object

There are strict conventions within the conservation profession governing the presentation of objects. These are governed by ethical considerations. The most important rule is that the original object is sacrosanct and should remain so regardless of the interventions undertaken. Secondly, any addition should be reversible and should be obvious on scrutiny. And lastly, no interpretation should be imposed on the object. This approach would normally exclude reconstruction or restoration. By implication, it means that the only improvement possible is to remove the visual impact of disturbing damages.

Pl. XV   Uncovering with a tailor-made hardened-steel tool.

In the case of the Clare Island Abbey the presentation was not easy. On the exterior the thin lime render, or shelter coat, that was applied to the chancel has given rise to some unfavourable comment (Pl. XVI). It is, however, historically correct as any medieval building as important as the Abbey would have had a protective render. It was also an absolute necessity for the improvement of the interior climate.

On the interior probably fifty percent of the original painted surface was lost through salt eruptions and plaster loss. This made it almost impossible for anybody who was not very familiar with the object to read the images. It was finally decided to reconstruct the painted vault ribs where they were missing. Such reconstruction involved joining the dots, so to speak, between the extant paint traces, rather than pure conjecture (Pl. XVII, Fig. 1). None of the original ribs were overpainted. While some might consider this decision unethical, it was felt that the legibility of the painting was improved so dramatically that it could be justified and accepted. This reconstruction meant that the eye, which had previously been drawn to and distracted by the damaged areas, was free to observe the images themselves.

Pl. XVI Exterior view of the Abbey with shelter coat applied to the chancel.

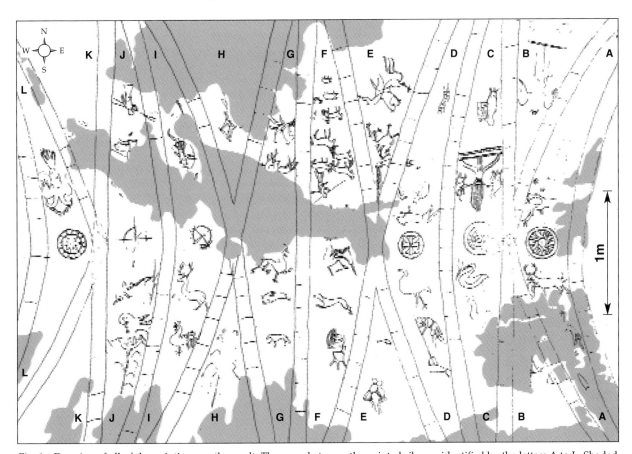

Fig. 1 Drawing of all of the paintings on the vault. The areas between the painted ribs are identified by the letters A to L. Shaded areas indicate plaster loss.

Acknowledgements

There was no prototype or precedent for the problems presented on Clare Island. I knew of no other cycle of wall paintings that had been reclaimed from such an advanced state of decay. The interpretation of the complex situation has evolved in close collaboration with numerous colleagues, many of whom helped on site and gave generously of their knowledge and expertise. The success of the project relied on the input from many specialists, all of whom are named in the conservation report. Because of the beauty of this special site and the uniqueness of the challenge, many offered their services free of charge, including the microbiologists and the radiocarbon dating experts.

In particular I would like to thank Madeleine Katkov, the co-director of the project; Karena Morton, for all her considerable efforts, not only in working with us on site, but also in producing this contribution, which would not have come together without her; and Conleth Manning, for his help in editing this paper and for his patience. I should like to thank the clients, The National Monuments Service, including both the architects at head office—Willy Cumming, who was there at the outset and clarified the initial approach, and more recently Paul McMahon. Thanks are also due to the staff at the Castlebar and Dromahair depots, who looked after our needs, in particular John Corcoran, Frank Reid and Kieran Reid.

Pl. XVII Presentation of the paintings: joining the dots between fragments of the painted ribs.

Last but not least I should like to thank the people of Clare Island for bearing with us through all the vicissitudes of a difficult and arduous task.

NOTES

1. This material had become available only a few years earlier. It had been tested by two major European conservation research institutions only the previous year (1995) and had been given a positive evaluation.

REFERENCES

Bläuer Böhm, C. 1994 Mineralogical examination. Unpublished report, Fachhochschule Köln, BMFT-Projekt.

Petersen, K. 1992 Microbiological examination, description of species, effects of UV-C irradiation. Unpublished report, Universität Oldenburg.

Raschle, P. 1996 Microbiological examination, sequence of colonisation, preventative measures. Unpublished report, EMPA Eidgenössische MaterialPrüfungsAnstalt.

Westropp, T.J. 1911 Clare Island Survey: history and archaeology. *Proceedings of the Royal Irish Academy* **31** (1911–15), section 1, part 2, 1–78.

CATALOGUE OF THE WALL PAINTINGS

Karena Morton and Christoph Oldenbourg

ABSTRACT

We present a detailed catalogue of both painting phases as they survive today. For each section and subsection of the wall paintings the description of the existing images is contrasted with the material presented by Westropp in the original Clare Island Survey. This is undertaken not only to highlight the new discoveries that were made during conservation but also to augment our knowledge of the paintings by drawing attention to some images that have been lost or have deteriorated since Westropp recorded them.

Introduction

A comparison of the images that were recorded by Westropp in 1909–11 and those that are currently visible is worth undertaking for a number of reasons. Firstly, some of the images visible in 1909–11 have since been concealed by particularly stubborn calcification and other mineral layers, which have yet to be carefully removed (Oldenbourg, this volume, pp 49–60). Secondly, and in contrast, many of the images recorded by Westropp are now, after conservation, much clearer, which helps identify the subject matter depicted. Thirdly, some of the images were lost between 1909–11, when Westropp recorded the paintings, and 1991, when conservation began. Lastly, many additional images have been revealed through conservation, which were, even at the time of the first survey, so severely deteriorated and masked by calcite, microbiology, salts and other products of decay that they were invisible or remained undetected. This fact is important as it indicates a time frame for the more recent history of decay at this site.

In his account, Westropp referred to the difficulty of describing certain colours that might have been faded or altered pigments (1911, 33). This difficulty is also referred to by Oldenbourg

(this volume, pp 49–50). For brevity, Westropp used the term 'blue' throughout. In the present account all nondescript colours are treated as one. Even analysis of every occurrence might not reveal adequate distinctions. Rather than using Westropp's term 'blue', the term 'of faded colour' has been adopted here to mean that no particular colour is suggested. It solely describes a colour of unidentified origin, such as lines attributable to decay products, mineral staining or faded original pigments. This term allows for differentiation from a hue clearly attributable to a decay product.

The authors have intentionally adopted Westropp's numbering system for the second phase of wall painting (vault area) in these articles as a way of acknowledging the effective and comprehensive work undertaken by him during the original Survey of Clare Island. However, it is noteworthy that the numbering system for the vault used by Westropp in the published survey is different from that used by him during his on-site documentation.[1]

CATALOGUE OF THE PHASE ONE WALL PAINTINGS

The Phase One paintings are to be found on the north and south walls, below the level of the

Fig. 1   Drawing of the north wall (traced from a photograph) showing both phases of painting.

Pl. I   Westropp's drawing of the north wall showing on the right the ragged opening into the attached domestic wing (RIA).

vault. When conservation began, microbiology and other products of decay disguised virtually all of the Phase One wall paintings.

North Wall (Fig. 1, Pl. I)

The north wall is divided equally into three sections. In the western section a few small traces of paint can be seen. A small opening or squint is located centrally in this section, and to the right—next to the tomb niche (or 'Easter Sepulchre' as Westropp called it)—is the O'Malley plaque. The deteriorated condition of this area means that it cannot be established with certainty whether this paint belongs to Phase One or Phase Two.

A tomb recess with carved tracery and a pinnacle on either side occupies the central section of the north wall. A distinct band of red paint is visible on the plaster that abuts the tomb spandrels and finials on either side (Pl. II).

In addition, a few small traces of red paint survive in the joints of the canopy itself. To the right (east) is a painted representation of the same type of tomb canopy. The painted version has been partially covered by the later addition of a load-bearing pier, which was inserted to support the construction of the stone vault (Pl. III).

In this eastern section the colours chosen for the painted tomb canopy provide a clue as to the colours used to embellish the carved version. Indeed, the remaining paint traces on the carved canopy tally with this suggestion. The use of white, red or unpainted (the black of the stone) for the different carved faces matches exactly the colours used on the painted version. The ogee frame and the three finials have white bevels, with the recesses on the wall picked out with a red line. The very front edge was left unpainted, retaining the natural stone colour of black. The 'painted tracery' was painted entirely in red, with

Pl. II   A section of the north wall of the chancel showing part of the tomb, with its painted counterpart to the right and a hound chasing a deer.

Pl. III   The painted counterpart of the tomb canopy at the east end of the north wall.

black lines indicating the joints in the stonework. This fact confirms that the tomb niche was incorporated as part of the overall decorative concept for the Phase One wall paintings. The uncovering of this painted version is a most exciting discovery and gives weight to the concept of integrated decorative schemes where every element is conceived as part of the same canvas, so to speak. It has long been debated whether carved stone features were left in their natural (and 'beautiful') state or, as in this case, were embellished with paint.

To the east of the carved canopy is a secondary aperture, which cuts through and has damaged the painted canopy and obviously post-dates the Phase One paintings. In the east spandrel of the sepulchre, between the masonry of the arch and pinnacle, is a red-outlined stag or buck with a slightly pinkish body and black antlers, running westwards (Fig. 1, Pl. II). To the east of the pinnacle a large red hound pursues this stag (Pls II, III). Further paint can be made out in the eastern spandrel under the Phase Two plaster layer (which bears the easternmost painted rib and corbel). These paint traces may originally have represented a huntsman accompanying the hound.

Westropp noted the existence of two painting phases in the area abutting the stonework of the 'Easter Sepulchre'—'The rough face was covered by a good undercoat, on which was painted an ornamental design in deep, rich crimson...' (1911, 32). He conceded that it was difficult to identify with certainty much of the earlier scheme. He does not comment on the stag, hound or painted canopy, and we must assume that none of this was visible to him.

South Wall (Fig. 2)
The only surviving image from the Phase One paintings is that of a figure of a horseman galloping eastwards, located at the west end of the south wall above the door leading to the stairs (Fig. 2). The horse (which is pony-sized in relation to the rider) wears a saddlecloth (of the same colour as the horseman's coat) extending over his croup, secured by two black girths, crupper and breast-girth. The snaffle bridle is also black. There is a strange curved black line in front of the horse's face that does not correspond to a chanfron: its role is unclear (Pl. IV).

The rider has a hauberk of small overlapping square pieces, painted black with details incised or sometimes outlined in red, probably with the

Fig. 2   Drawing of the south wall (traced from a photograph) showing both phases of painting.

Pl. IV The mounted soldier with a spear, which appears above the doorway on the south wall. This is part of the Phase One painting scheme. Plaster associated with the later doorway can be seen below the horse's feet. A remnant of Phase Two plaster and painting can be seen above the horse's hindquarters.

fringe of the garment worn underneath showing at its lower edge. His arms are protected by sleeves of mail, and he wears a simple conical red helmet without a nosepiece. His face is also black. Above his head he wields a long spear or javelin. There is a further fragment of a black line, at a slightly different angle. It is difficult to establish if this line is part of his weapon or that of an opponent to the east, now lost. There is no indication of a saddle or stirrups. No trace of the horseman was visible to Westropp.

Westropp interpreted the doorway to the south stair as being an original feature and therefore ascribed the decoration on the soffit ('red bands to either side, with a saltier between') as belonging to the earlier painting phase. The plaster on the stone arch of the door is proud and partially overlaps the horseman. The arch must therefore post-date the Phase One paintings (Westropp 1911, 32, and Morton, this volume, pp 97–121).

CATALOGUE OF THE PHASE TWO WALL PAINTINGS

The Vault (Pls V, VI, VII, Fig. 3)
'Architectural framework'

The vault is divided into four bays by five false ribs, painted to imitate stone, that run north–south. A decorative corbel terminates each rib on either side of the vault. Each of the four bays is further divided into four sections by diagonal ribs, with a lozenge at the apex where the diagonals cross. All the ribs are painted black and set out with deeply incised lines. Further incised lines divide the ribs into 'masonry blocks'.

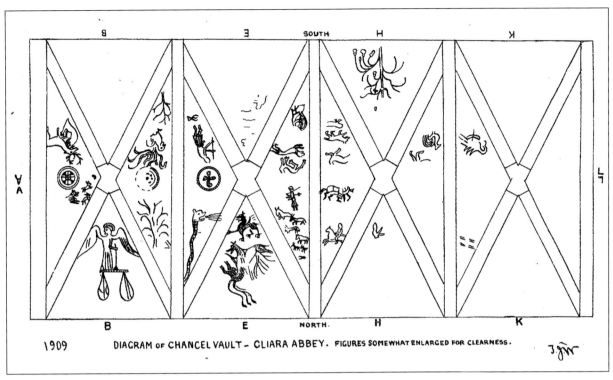

Pl. V   Westropp's drawing of the paintings on the vault as published in the original Clare Island Survey report.

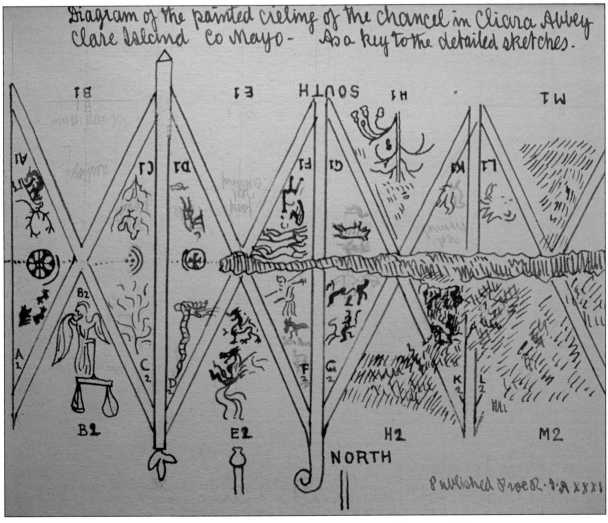

Pl. VI   Westropp's drawing of the paintings on the vault from his notebooks (RIA MS 3.A.52, vol. 7).

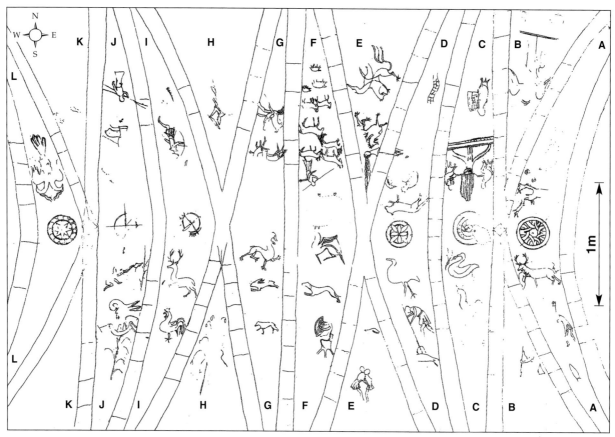

Fig. 3 Drawing of all of the paintings on the vault as traced from the photograph reproduced as Pl. VII. The areas between the painted ribs are identified by the letters A to L.

Pl. VII Overall view of the paintings on the vault.

The corbels

The design of each corbel seems to be different (See Figs 1, 2; Pls I, VIII). On the north wall the designs of Corbels 1 and 5 are too damaged to decipher. Corbel 2 has three leaves, Corbel 3 has a volute ending with three small leaves[2] and Corbel 4 has a fragmentary eastward curl, adjacent to the east pinnacle of the sepulchre. On the south wall Corbel 1 is very deteriorated, but three small leaves can be made out and there may well have been a volute similar to that on Corbel 3 on the north wall. Of Corbel 2, only a few small scraps of black survive. Corbels 3 and 5, however, have perished. The equilateral triangle on the 'middle' corbel described by Westropp[3] is no longer visible (1911, 33). Corbel 4, uncovered during conservation, is shaped like a stylised tree.

Pl. VIII Terminals of painted vault ribs as sketched by Westropp (RIA).

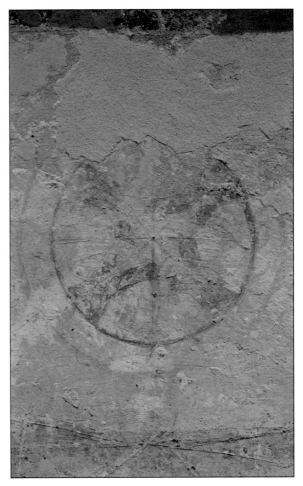

Pl. IX Boss A.

Decorative bosses

There are six bosses at the apex of the ceiling. These are found in Sections A, C, D, I, J and L. They were set out with incised lines using compass and ruler.

Boss A: (Pl. IX) Nearly complete. There are two concentric circles: the outer one is yellow, the inner one red. Red quarter-circles form a diamond shape in the inner red circle. At the points of intersection, yellow semicircular bands are interlaced.

Boss C: (Pl. X) More than half of this boss survives next to a large hole in the vault. There is a small red cross in the centre surrounded by four concentric circles: the first two red, the next yellow and the outermost red. The innermost circle is filled with yellow dashes and the next with red dots and faded dashes. It is of interest that the eastern edge of the hole in the vault is polished as though by a bell rope.

Pl. X   Boss C. The hole was probably for a bell-rope as there is some polishing of the plaster consistent with such a use.

Pl. XII   Boss I.

Pl. XI   Boss D.

Pl. XIII   Boss J. The pink hue at the top and bottom of the image should be disregarded — it is due to a processing error.

Boss D: (Pl. XI) Complete. A cross of eight arms, alternately red with rounded ends and of faded colour with pointed ends, is set within two concentric circles. The inner circle is yellow and is overlapped by the points of the faded-coloured arms. The outer circle is red. There is evidence of a background colour that differs from the faded grey–brown of the pointed arms.

Boss I: (Pl. XII) A cross of eight pointed leaves is set in the centre of concentric red and yellow circles. The leaves are alternately red and yellow and extend to the outer circle. Beyond this are radiating dashes of red within an outer circle of faded colour.

Boss J: (Pl. XIII) A damaged red circle and radiating lines survive among the incised setting-out lines.

Pl. XIV   Boss L.

Boss L: (Pl. XIV) A pointed cross of eight arms, alternately yellow (with a red outline) and red, is set within a border of curved red dashes within a red circle.

Westropp recorded the western three bosses only, and he was not able to observe even these at all well. The three at the eastern end were obviously completely concealed by microbiology and other deposits.

The Painted Images
The painted scenes loosely follow the incised lines applied when mapping out the overall design (see Oldenbourg, this volume, p. 52).

The north side (see fold-out illustration 1)
Section A
Below Boss A, at the top, is a prick-eared yellow hound or retriever chasing three hares with long hind legs (Pl. XV, Fig. 4). The upper hare is yellow, the middle one incised, and the lower one red. They run eastwards. The scene below is now very damaged. It appears, however, that a man in red has just opened his hand to release a red hawk, which flies after the hares. The lower plaster is much decayed. This must be a falconry scene. Only the yellow hound and the yellow and red hares were visible to Westropp (Pl. XVI).

Section B
St Michael with his scales of judgement occupies this section (Pl. XVII). This area, even after conservation, remains heavily disguised by a dark grey calcification. However, the incised lines indicate the outline of the scene.

Current research may provide new treatment techniques and methods to safely uncover this area: such techniques were not available during the recent conservation programme. For the moment we rely on Westropp's comments and drawing of this area as reproduced in Pl. XVIII (1911, 34). However, it does appear from Westropp's descriptions that this area was already affected by calcification, obscuring the details of the image.

Pl. XV   Hound chasing hares.

Pl. XVI   Westropp's illustration of the hound chasing the hares. The drawing to the right is masked because it does not appear in this position on the ceiling (RIA).

Fig. 4   Scribing of hound chasing hares.

Pl. XVII   Present state of the painting of St Michael obscured by later calcification.

Pl. XVIII   Westropp's illustration of St Michael as recorded by him in 1909 or 1910 during the original Clare Island Survey (RIA).

Pl. XIX Organ, with person on left playing and person on right working the bellows.

Fig. 5 Scribing of organ: the person on left plays the instrument while the person on right is working the bellows.

Section C

Below Boss C, at the top of this section, is a large organ on a red stand with splayed legs (Pl. XIX, Fig. 5). The organ has ten pipes, alternately red and yellow. To the west, a man in red on a yellow seat plays the organ. A second figure in yellow outlined with red sits on a red seat and works the (incised) bellows.

Further down, a yellow man on a red seat plays a lyre, as identified by A. Buckley (pers. comm.) (Pl. XX, Fig. 6). He faces east and has one hand on each side of the strings: his left hand is clearly visible behind the incised strings. The strings are attached to a tailpiece lying over a soundbox at the bottom of the instrument. Westropp described 'yellow branches and fruit' (1911, 34). He was clearly unable to make out these images as he described the background as being faded blue.

Section D

Below Boss C, at the top of this section, an incised lynx or wildcat walks eastwards (Pl. XXI, Fig. 7). Below, also incised, is a pale yellow cat with what appears to be a bird, the details of which are unclear.

Rising from the bottom of this section is a tall chequered 'serpent' standing on its tail (Pl. XXII; Fig. 8). It has a red eye and an opened yellow-striped snout. It spits a series of lines—possibly flames—across the rib towards the dragon in the bay to the east. In this section only the serpent was visible to Westropp (Pls V, VI), and he described it as 'divided into squares of red and blue...'. In fact, its body is made up with two rows of squares coloured red, yellow and white.

Section E

At the top of this section flames rise from a hand or a yellow gauntlet outlined in red (Pl. XXIII, Fig. 9). Just below, a winged red dragon with a spiky mane and four four-clawed feet faces the snake to the west (Pls XXIV, XXV).

Below is a puzzling construct of four animals (dogs or dragons?), with those at the sides back-to-back, and those at top and bottom facing (Pl. XXVI). They each share a head and forelegs with one neighbour and hindquarters with the other neighbour (Pl. XXVII, Fig. 10). Yellow outlines cross the red and emphasise the linking of the four bodies.

Pl. XX   Person playing a lyre.

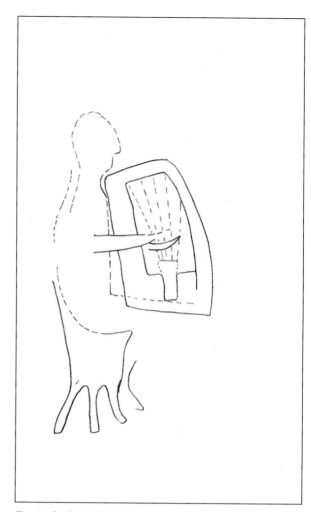

Fig. 6   Scribing of person playing a lyre.

Pl. XXI   Lynx or wildcat.

Fig. 7   Scribing of lynx or wildcat.

Pl. XXII   Fire-breathing serpent.

Fig. 8   Scribing of fire-breathing serpent.

Proceeding clockwise, the upper beast has red foreparts, a yellow belly, red hindquarters and a short yellow tail. The one to the east has yellow foreparts and a yellow belly and shares the same red hindquarters and short yellow tail. The lower beast has yellow foreparts and red hindquarters, with a long red-stranded tail. The beast to the west shares red foreparts with the upper beast and the long-tailed red hindquarters of the lower beast.

Much of this section was visible to Westropp, and he clearly studied it closely (Pl. XXVII). He described the red dragon as having 'a wolf's head, lion's body, and eagle's claws, with perhaps a trace of wings' (Westropp 1911, 34). Of the four

beasts, Westropp was only able to make out the red elements of the design. From these he ingeniously determined that the artist had changed his mind: 'The artist scribed a wolf at full gallop, with a hound swinging from its throat; afterwards he (or the painter if two were at work) retained the wolf's head, turned its body into a wing, and painted the dog so as to form the body, adding tails and paws as required' (Westropp 1911, 34–5).

Sections F, G, H, I, J and possibly K

We suggest that Sections F, G, H, I, J and possibly K form part of one scene, that of a cattle raid.

Pl. XXIII   Hand or gauntlet with flaming sword.

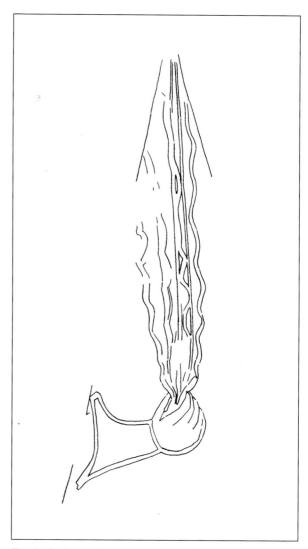

Fig. 9   Scribing of hand or gauntlet with flaming sword.

Pl. XXIV   Winged red dragon.

Pl. XXV   Westropp's illustration of the red dragon (RIA).

Pl. XXVI   Body-sharing animals.

Fig. 10   Scribing of body-sharing animals.

Pl. XXVII   Westropp's illustration of body-sharing animals (RIA).

Section F

At the very top is a large area of plaster loss; on the bottom edge of this two bright yellow clawed feet survive, and on its upper edge are two yellow tips of what might have been wings or ears.

Below this are a herdsman and his animals. He is dressed in a yellow cap and tunic; both are outlined in red. His left sleeve is painted red. He wears hose of faded colour and carries in his right hand a long pole, indicated by a single incised line, and in his left hand a faded shortbow. A quiver or scabbard of faded colour hangs from behind his waist. A yellow heifer leads the herd westwards, closely followed by a red heifer with yellow ears. Below these, further down the wall, is another red heifer, followed directly by an incised heifer of no distinct colour—possibly a pentimento. Just below this is a further yellow heifer (Pl. XXVIII). Further below, as the space between the ribs narrows downwards, two yellow sheep or goats appear—a horned yellow one following a red one.

At the narrow point of this section are pigs— a large one of faded colour with an incised outline (Pl. XXIX), followed by a small red one, with a little yellow one underneath. The eyes of several beasts were cut into the dried plaster at some point after the paint was applied. An opaque grey layer of calcite obscures much of this lower area.

Westropp described most of the elements in this section (Pls XXX, XXXI). However, some of the details must have been unclear at this time. Although he noted the herdsman's 'red staff, ending in a triangle, under his arm', he was not able to make out that this is an arrow lodged in the herdsman's back at the level of his shoulder blades, shot by one of the bowmen in Section I to the east.

Pl. XXVIII Cattle and armed herdsman with arrow in his back. This has been interpreted as part of a cattle raid.

Pl. XXX Westropp's illustration of the cattle and herdsman (RIA).

Pl. XXIX Goats and pigs.

Pl. XXXI Westropp's illustration of part of the cattle herd and the goats and pig (RIA).

Section G

Approximately twenty-five per cent of this section (from the apex down) is an area of plaster loss, but the lower half of a horseman survives. The figure wears red hose, and the faded colour above his knee may be a tunic (Pls XXXII, XXXIII; Fig. 11). He rides on a red pony with a yellow saddle blanket and has turned to face the bowmen in Section I. His pony has a yellow arrow in the chest. Interestingly, Westropp recorded a man standing before this horse and described this figure in detail: 'A man stands before it; he has red hose, and a "blue" tunic, with fringe and belt...' (Westropp 1911, 35). No trace of

the figure as described survives today: indeed, Westropp did not include it in either the overall drawing or the detailed sketches of the vault paintings. In addition, the man mounted on the horse in question was not described by Westropp and presumably was not visible. There is some confusion in the numbering of the different sections at this point in Westropp's article, which makes it impossible to interpret where exactly Westropp located the figure in question. However, given the detail of the description provided it seems likely that this figure was represented in the immediate area of the horse in question[4].

Pl. XXXII   Horse riders (part of the assumed cattle raid).

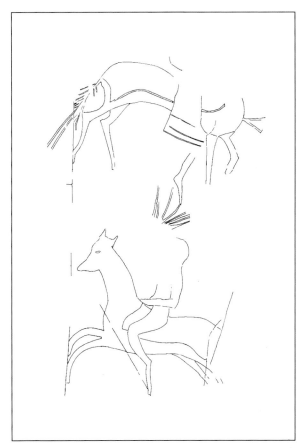

Fig. 11   Scribing of horse riders.

Pl. XXXIII   Westropp's illustration of the horse riders (RIA).

Below, a second horseman dressed in a red, close-fitting garment rides a yellow pony equipped with a saddlecloth and follows the animals in Section F to the west (Pl. XXXIII) Between these two horsemen are the fragments of a few red strands, which could possibly be feathers from headgear or flames.

Westropp noted the 'blue' of the saddlecloth and tunic of the upper horseman. There is plaster loss at the bottom of this section.

Section H

On the small area of plaster remaining in this section, a man in a conical helmet carries a weapon of uncertain type in his raised hand (Pl. XXXIV). The whole figure is painted in red. Above are scraps of yellow adjacent to a broken plaster edge, perhaps part of the 'yellow animal' that was already fragmentary when Westropp described it in 1909.

Sections I, J, K and L

Sections I, J, K and L[5] were not visible to Westropp during his visits as they were 'destroyed, save part of the ribs, by green growth' (Westropp 1911, 35). It is of interest to note that sections were also concealed by an overlying limewash (see Oldenbourg, this volume, p. 57) in addition to the 'green growth'.

Pl. XXXV   Bowman (part of the assumed cattle raid).

Pl. XXXIV Soldier with conical helmet with sword (part of assumed cattle raid). The inset shows Westropp's illustration of the same image (RIA).

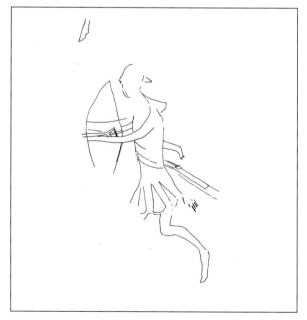

Fig. 12   Scribing of bowman.

Pl. XXXVI   Fragment of a horseman above a man carrying two spears (part of assumed cattle raid).

Fig. 13   Scribing of man carrying two spears.

Section I (mislabelled K by Westropp 1911, 35)
Below Boss I at the top of this section is an area of plaster loss. At the very top, adjacent to a large area of missing plaster, is a red horse's hoof. Below, a bowman moves rapidly to the west while in the act of discharging an arrow (Pl. XXXV, Fig. 12). He is dressed in a striped red and yellow kilt with red shirt and hose. Below this figure are the fragmentary remains of a second bowman. Both bows appear to be shortbows.

Section J (mislabelled L by Westropp 1911, 35)
Below Boss J the plaster is damaged, and much is missing at the top. The hindquarters of a red horse can just be made out. Below is a fragment of a second horseman. Dressed in a red tunic, he carries an incised spear. He rides westwards on a pony (of faded colour) with a yellow saddlecloth and an incised straggly tail. A curious red curl may be all that remains of hose or a shoe. Below is a yellow man in a tunic with a red belt and hose of faded colour, carrying two long red spears whose deeply incised ends extend onto the rib to the west (Pl. XXXVI, Fig. 13).

Section K (mislabelled M by Westropp 1911, 35)
The remains of some incised lines with red and yellow colouring appear near the plaster loss at the top of this section. The design remains unclear.

Section L (mislabelled N by Westropp 1911, 35)
Below Boss L at the top of this section is a double-headed eagle. Its body and heads are red, its tail and pinions are outlined in red, and red flames issue from the heads (Pl. XXXVII, Fig. 14). Nothing further can be made out.

The south side (see fold-out illustration 2)
Section A
Boss A: At the top of this section a yellow stag with red antlers faces east. Three hounds are attacking him (Pl. XXXVIII, Fig. 15). The most obvious, a red hound (or wolf), is at his throat, while an incised hound can just be made out above and behind him. The third hound is red and has its paws against the stag's hind legs. Westropp noted the stag, with just one red 'wolf' (Pl. XXXIX). The lower plaster is damaged. Fragments of a fisherman like that in Section D, together with the fish he caught, survive.

Pl. XXXVII   Double-headed eagle.

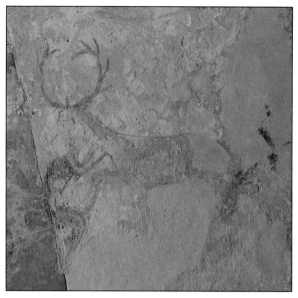

Pl. XXXVIII   Stag attacked by hounds.

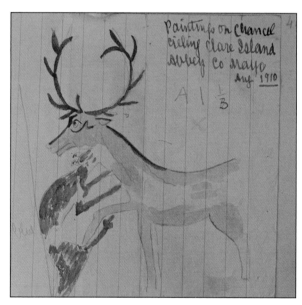

Pl. XXXIX   Westropp's illustration of the stag attacked by hounds (RIA).

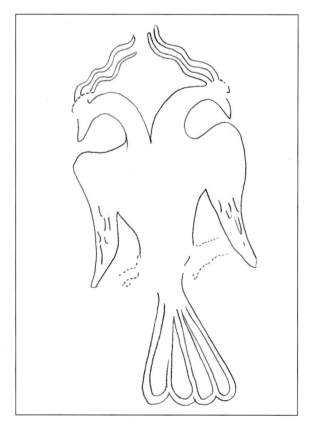

Fig. 14   Scribing of double-headed eagle.

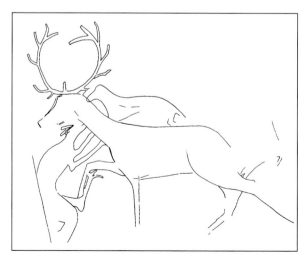

Fig. 15   Scribing of stag attacked by hounds.

Section B

All that survives is a fragment of a red man with yellow sleeves plying a long pole, possibly a harpoon or spear (Pl. XL, Fig. 16). The surface has mostly gone; it was defaced already when Westropp recorded it.

Section C

Boss C: Below Boss C an incised square-beaked swan faces east and is apparently feeding on a red plant. The swan may well be swimming, as its legs are not shown (Pl. XLI, Fig. 17). There is further red paint, possibly a second plant, between the swan and the cock below. The cock is red and struts eastwards with upraised claw. It has a large tail and red wattles, but its comb is lightly incised only. Further below the plaster is much damaged, but there are the remains of other incised lines. Westropp noted a large red cock and floriated yellow and red sprays (Pl. XLII).

Section D

Boss D: Below Boss D is a large yellow bird—a heron or crane—facing east (Pl. XLIII, Fig. 18).

Below this is a fisherman who has speared a fish using a trident (Pl. XLIV, Fig. 19). The figure,

facing east, has an incised outline, and the greater part of his tunic is of faded colour with a red outline down the back. The left leg of his hose is yellow and the right leg and his sleeves are red, while his kilt is red and yellow.

Westropp described the hose as having 'red knee-caps' but these are not obvious now. He also described the tunic as being 'held in by a waist-sash, ending in streamers and a red bow' (1911, 35), but again this is not currently visible (Pl. XLV).

Section E

At the top of this section is the incised body of a dog of faded colour. Although the head is no longer visible (due to plaster loss), a red collar can be made out. The dog faces east (Pl. XLVI, Fig. 20). Below this are two embracing or wrestling figures (Pl. XLVII, Fig. 21). The one on the left is painted all in red, and the folds in his skirt are indicated. The figure on the right is painted all in yellow, but the details of his dress are unclear. Nothing further has been made out in this section, which supports much calcification. Westropp could only discern 'faint red animals overhead, and a red man leaning over a yellow object' (1911, 35).

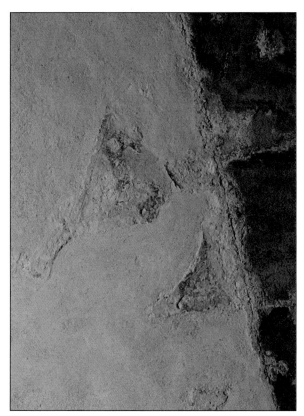

Pl. XL Fragmentary painting of man fishing with trident.

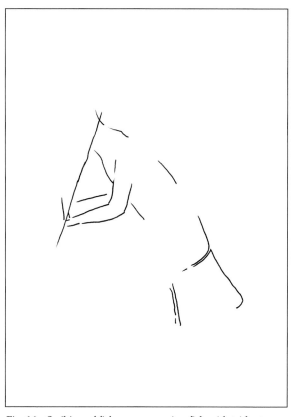

Fig. 16 Scribing of fisherman spearing fish with trident.

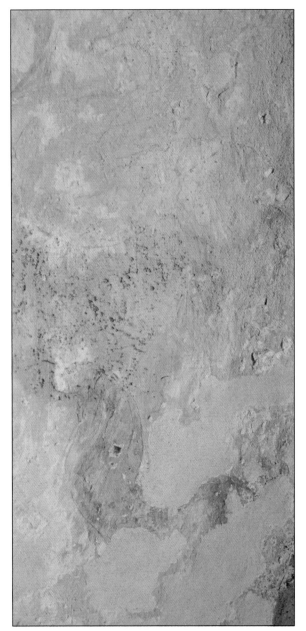

Pl. XLI   Swan feeding on plant above cock.

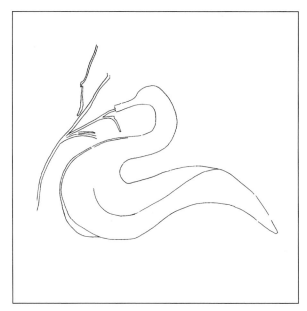

Fig. 17   Scribing of swan feeding on a plant.

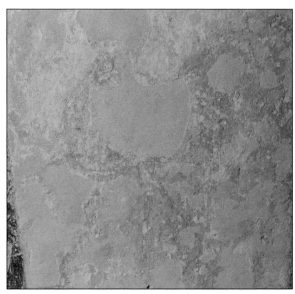

Pl. XLIII   Heron or crane.

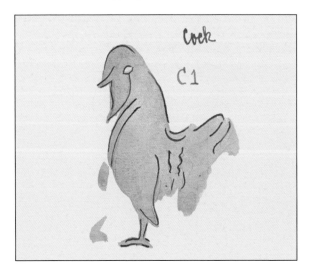

Pl. XLII   Westropp's illustration of the cock (RIA).

Fig. 18   Scribing of heron or crane.

Pl. XLIV   Fisherman spearing fish with trident.

Fig. 19   Scribing of fisherman spearing fish with trident.

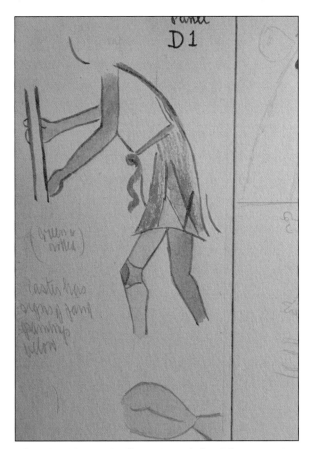

Pl. XLV   Westropp's illustration of the fisherman using trident (RIA).

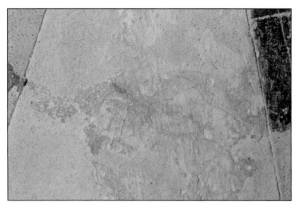

Pl. XLVI Small hound running.

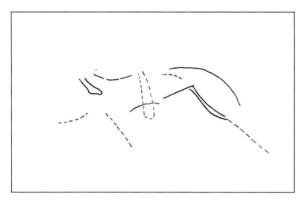

Fig. 20 Scribing of small hound running.

Section F

Extending to the apex at the top of this section is a winged dragon or griffin of faded colour (Pl. XLIII, Fig. 22). It has a slender leonine body with wings, a tail and clawed feet. The creature faces the yellow dragon in the next bay to the east. Below it is a red greyhound pursuing the hare in the bay to the east (Pls XLIX, L).

At the bottom is an incised harpist in robes of now faded colour but with some traces of yellow and red sleeves. He sits on a red chair with a yellow seat. His harp has a red frame relieved by a yellow line (Pl. LI, Fig. 23). Nothing further has been uncovered in this section.

Westropp stated that the harpist was 'in long yellow robes, holding a characteristic Irish harp, with six red strings' (1911, 35) (Pl. LII).

Section G

At the top, again at the very apex, is a fabulous yellow beast or dragon that walks westwards, facing the griffin in the adjacent bay. It has a spiky mane and clawed feet, like the red dragon in Section E on the north side, but no wings are

Pl. XLVII Wrestlers.

Fig. 21 Scribing of wrestlers.

Pl. XLVIII   Winged dragon or griffin.

Fig. 22   Scribing of winged dragon or griffin.

Pl. XLIX   Running hound.

Pl. L   Westropp's illustration of the running hound (RIA).

visible (Pl. LIII, Fig. 24). Below this, a yellow hare runs eastwards away from the hound in the next bay (Pl. LIV).

At the bottom, walking eastwards, is a red animal—probably a fox (Pl. LV). Westropp described this animal as 'a very natural prowling red wolf' (1911, 35) (Pl. LVI). Westropp also made out two yellow animals, which he described as two dogs. Nothing further can be made out in this section.

Section H
The only image surviving in this section is of a yellow tree with slightly dangling flowers or fruit, climbed by an incised animal whose shape suggests a cat (Pls LVII, LVIII). Westropp only recorded 'a yellow tree, with pear-like fruits' (1911, 35).

Section I (mislabelled K by Westropp 1911, 35)
Under Boss I at the top is a halting stag, painted in yellow (Pl. LIX, Fig. 25). The stag is part of a hunting scene that spans Sections I and J. Below is a red cock with a red and yellow tail, red comb,

and yellow beak and wattles (Pls LX, LXI; Fig. 26). There are some unclear incised lines below. Westropp noted 'a red hound, a yellow animal, and a faint trace of some red figure...' (1911, 35). The latter he recognised on closer examination as a bird ('like a cock'). There is now no trace of the red hound described by Westropp.

Section J (mislabelled L by Westropp 1911, 35)
Under Boss J at the top of this section a red man crouches (Pl. LXII, Fig. 27). He restrains a hound (of faded colour with incised outline) on a lead. The slipper and hound face the stag in Section I. The leash, a slip lead, is doubled through the hound's red collar so that it can be released instantly. There is a yellow tree between the hunters and the stag.

Further down is a large yellow pelican outlined in red (Pl. LXIII, Fig. 28). Pecking at her breast, she feeds her young—traces of red (blood) can be made out. The young birds are of faded colour but have incised outlines: they sit in a red nest. The nest resembles a boat in shape and is supported on a yellow tree. Little of this image

Pl. LI   Harpist.

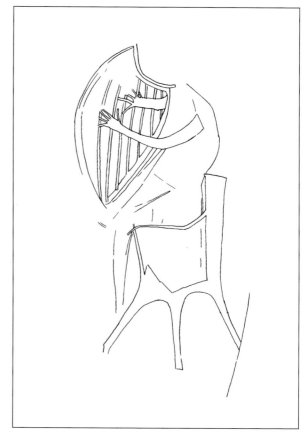

Fig. 23   Scribing of harpist.

Pl. LII   Westropp's illustration of the harpist (RIA).

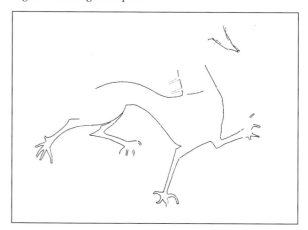

Fig. 24   Scribing of fabulous yellow beast or dragon.

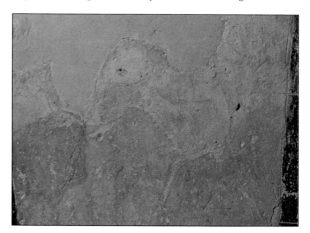

Pl. LIII   Fabulous yellow beast or dragon.

Pl. LIV   Hare.

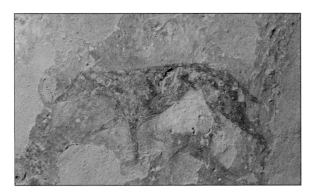

Pl. LV   Hound (possibly a fox).

Pl. LVI   Westropp's illustration of the hound (possibly a fox) (RIA).

Pl. LVII   Climbing animal (perhaps a cat) and tree.

Pl. LVIII   Westropp's illustration of the tree (RIA).

Pl. LIX   Halting stag.

Fig. 25   Scribing of halting stag.

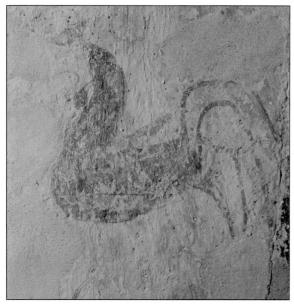

Pl. LX   Cock.

Fig. 26   Scribing of cock.

Pl. LXI   Westropp's illustration of the cock (RIA).

Pl. LXII   Man with hound on slip-lead.

Fig. 27   Scribing of man with hound on slip lead.

Pl. LXIII   Pelican.

Fig. 28   Scribing of pelican.

Pl. LXIV   East wall of chancel with fragments of Phase Two paintings.

was clear to Westropp, but he correctly worked out the subject matter.

Section K (mislabelled M by Westropp 1911, 35)
The plaster surface is eroded. Only scraps of red and yellow and curved incised lines survive. No image can be made out.

Section L (mislabelled N by Westropp 1911, 35)
Under Boss L the plaster surface is also eroded. Only a scrap of yellow survives. No image can be made out.

The walls
The west wall
Although there is no paint distinguishable on this wall, the plaster remains overlap the masonry of the chancel arch and are presumably contemporary with the Phase Two painting scheme.

The east wall (Pl. LXIV)
In front of the east wall is an altar table that is joined to the north wall by a narrower sill. Above the altar in the centre of the wall is the east window, set into a deep reveal with two cusped ogee-headed lights. A lot of plaster loss has occurred on this wall, and the surviving paint is almost entirely yellow, with just a few scraps of red to the south side (Pls LXIV, LXV, Fig. 29). A relieving arch, creating a shallow frame, spans a pier in the north and south corners. Small areas of the painted features can be made out, however. The shallow outer arch seems to have been decorated to resemble carved mouldings, with a broad band of lines painted like 'barley sugar' twists. A painted *trompe l'œil* gable surmounts the window reveal. There are further painted gables over *trompe l'œil* niches on either side of the window. There is painted tracery or scrollwork within the triangular sections of the gables. The reveals of the window recess were also decorated. On the south of the splay of the window a series of incised lines may have been a crucifixion scene, while on the north splay painted and incised floral motifs can be made out. It has not been possible to decipher the design on the east wall any further as this area has suffered so extensively from plaster loss.

Fig. 29   Drawing (traced from a photograph) of the east wall and altar showing traces of painting from Phase Two.

Pl. LXV   Westropp's drawing of a detail on the east wall (RIA).

The north wall
Placed in the centre of the back wall, within the tomb niche, is the upper part of Christ on the cross (Pls LXVI, LXVII). His head is missing, but the hands and arms are decipherable. The cross is red. To the west are traces of a red-robed figure; the remains of a figure to the east are incised only. This part of the painting is considered to be Phase Two on account of the similarities in the mortar properties noted during conservation (Oldenbourg,

Pl. LXVI   Crucifixion under canopy of tomb (RIA).

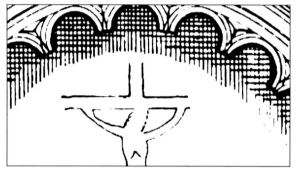

Pl. LXVII   Westropp illustration under canopy of tomb (RIA).

this volume, p. 51) and because the figures are of the same, smaller, scale as those on the vault (Morton, this volume, p. 98).

There are a few small traces of paint to be seen to the west of the tomb niche. As mentioned above, it is unclear whether these traces belong to Phase One or Phase Two.

The south wall

On the soffit of the stairway arch the orange–red edging bands and central saltire with incised setting-out lines are still visible. Above the piscina and on its return in the adjacent window-splay are more fragments of the same orange iron oxide, with incised lines. There seems to have been a simulated cusped canopy. Numerous fragments of incised lines survive on both reveals. Westropp misinterpreted both these fragments as belonging to the first scheme. Further scraps of red and yellow can be seen on the opposite return of the same window.

Acknowledgements

The authors would like to acknowledge the contribution of the late Anna Hulbert (Pl. LXVIII), conservator and art historian, who carried out most of the early work on cataloguing the wall paintings during the 1997 conservation season. Many of her observations are included in this catalogue. Many of T.J. Westropp's fine drawings and watercolours have been reproduced in this paper by kind permission of the Royal Irish Academy. These illustrations come from one of Westropp's many notebooks of sketches undertaken during his working life, now held in the Royal Irish Academy (RIA 3A.52). In this sketchbook some pages contain several images from different parts of the vault. With the exception of Figs 1–3, the illustrations are based on drawings A. Murray, undertaken in the last years of the project. Some of these have been modified by Christoph Oldenbourg in the preparation of this paper. We are very grateful for all the editorial work undertaken in the preparation of this paper, in particular on all the illustrations and photographs. We would like to thank Professors Roger Stalley and John Waddell and especially Conleth Manning.

Pl. LXVIII   The late Anna Hulbert in front of the 'cattle raid', cataloguing the wall paintings on the vault at Clare Island Abbey.

NOTES

1. In Westropp's original sketch of the vault paintings he used the letters A–H and K–N to designate the different areas between the painted ribs. He later changed this system to include the letters I and J (thereby making letters M and N redundant) when preparing his work for publication (Westropp 1911, pl. IV). However, when he described particular scenes in detail in this publication he must have referred to his original sketches and made the mistake of not adjusting the corresponding letters in line with the change made to the overall sketch of the vault paintings.

2. Westropp (1911–15, 33) described only the three central corbels on each side of the chancel. Hence, under the present numbering system, Corbel 2 corresponds with Westropp's Corbel 1, Corbel 3 corresponds with Westropp's Corbel 2 and Corbel 4 corresponds with Westropp's Corbel 3.

3. The 'middle' corbel is Corbel 3 on the south side under the present numbering system.

4. On examination of the space available in front of the horse in Section G it does not seem possible that such a figure could have fitted before the horse and certainly there is no indication of a man in front of the horse in Westropp's sketch of the area (Pl. XXXIII). However, there is substantial plaster loss above the horse. Perhaps the standing figure as described was actually positioned above and slightly to the right of the horse, thereby appearing to be in front of it. Westropp's description does, however, indicate that there has been a sizeable plaster loss in this general area since the 1909–11 survey.

5. The images described in Sections I, J, K and L were mislabelled by Westropp (1911–15, 35) as K, L, M and N, as he overlooked his own amendment to the overall numbering of the vault sections when preparing the drawing for publication.

REFERENCES

Westropp, T.J. 1911 Clare Island Survey: history and archaeology. *Proceedings of the Royal Irish Academy* **31** (1911–15), section 1, part 2, 1–78.

ICONOGRAPHY AND DATING OF THE WALL PAINTINGS

Karena Morton

ABSTRACT

The Abbey on Clare Island contains the most extensive array of medieval wall paintings discovered in Ireland to date. The images seem to occupy a space somewhere between the secular and religious and between the real and imaginary worlds. The iconography of the images is discussed both individually and in groups, where a number of images appear to relate to each other, and the question is posed as to whether there is an overall theme. Although individual scenes can be found in other church wall paintings, the second phase of painting at Clare Island is so far unique in the range of its motifs and the manner of their depiction. It is difficult to find contemporary parallels for all the images even in other media.

Introduction

The Abbey on Clare Island contains the most fascinating sequence of medieval wall paintings discovered in Ireland to date. Since the number of sites in Ireland with surviving wall paintings from the medieval period is small and the number of sites with more than one decorative phase is even smaller, it is remarkable that a church of this size and location was so comprehensively decorated not once, but twice. It is also remarkable that so much of both schemes of painting survived in these exposed and saline conditions. Following the conservation programme commissioned by the National Monuments Service and directed by conservators Madeleine Katkov and Christoph Oldenbourg, enough of the two painting phases survives to warrant a closer look, although neither phase is complete. Certainly, the Phase Two painting has the largest repertoire of motifs known from a single Irish wall painting of late medieval date. This painting finds its closest parallels not in wall paintings, or indeed in woodcarvings, tapestries or manuscripts, but in stone carvings such as those on the west doorway

at Clontuskert Priory, Co. Galway. The Clare Island paintings offer a comprehensive iconographic programme that ranks with the cloister arcade at Jerpoint Abbey in terms of the number and range of motifs. However, the relative lack of comparable material in Ireland makes it difficult to study the Clare Island images.

The details of the two painting phases have been described extensively elsewhere in this volume (Oldenbourg, pp 49–60; Morton and Oldenbourg, pp 61–95). In broad terms the materials and technologies used in both wall painting phases are the same. In addition to the clear stratigraphic distinction between the two painting phases, with the Phase Two painting on an overlying plaster layer, there are a few other ways of distinguishing the two phases. In the earlier (Phase One) painting, the figurative images were first scribed into the plaster. Colour was applied either in bold outline strokes with the interior of the image left unpainted, or the image was filled with block colours with few additional details. The colour palette is limited to white, red and black. In the Phase Two paintings the range of images is far more extensive, and their scale is considerably smaller than those of Phase One. Again, the design was mapped out with lines incised into the plaster and the paint was applied subsequently. On this painting there is a broader palette, with white, black, red, yellow and one, or perhaps two, now faded colours, and there is a greater fluidity to the quality of the painting.

Phase One wall painting

The Phase One painting is found on the north and south walls of the chancel. The north wall is dominated by a canopied tomb, which is roughly centrally placed. A tall ogee-headed arch with openwork tracery and tall pinnacles surmounts the tomb niche.

Remnants of paint on the canopy were uncovered during conservation along with a copy of this canopy feature that had been painted on the plaster to the right of the tomb (Morton and Oldenbourg, this volume, pp 63–65). The embellishment of carved elements with colour has precedents elsewhere: for example colour was an integral part of wall painting decoration at Abbeyknockmoy, where a pattern of black and white chevrons survives under the head of John the Baptist at the apex of the O'Kelly tomb (Morton 2004, 314). At Corcomroe Abbey the tomb of Conor na Siudaine O'Brien still retains a band of red paint on the edge of the stone canopy (There are several instances of painted stone details from both the *in situ* remains and excavated fragments from Kells Priory, Co. Kilkenny (Morton, forthcoming). It comes as no surprise that the tomb canopy at Clare Island Abbey was painted, as it is clearly the centrepiece and an essential component of this phase of interior embellishment.

Only two figurative vignettes survive from the Phase One painting, and neither of these has an obvious religious meaning. On the north side is a hunting scene—a stag pursued by a hound is positioned between the pinnacles of the canopy (Pl. I). It is of interest to note that while the stag is painted with assured brush strokes, it deviates from the intended posture as indicated by the outline scribed into the wet plaster (Oldenbourg, this volume, pp 51–52). Clearly, the artist originally intended the stag to be standing rather than fleeing. Further to the east, behind the

Pl. I Phase One, north wall: stag hunt with a hound in pursuit of a stag.

hound, are some tiny traces of paint, which may have been part of an image of a huntsman or another member of the hunting party. Stag hunting also features in the later Phase Two paintings at Clare Island. However, the Phase One image differs from those represented on the vault and has no direct parallel in an Irish wall painting. Here the hound chases after the stag but has not yet caught him. On the later vault paintings hounds are shown either approaching with a huntsman from the front or in the act of attacking the stag at the neck and hindquarters. It is possible that the significance differs slightly in each representation. Perhaps the most similar scene in terms of composition is the wall painting from the *capella ante portas* at the Cistercian abbey at Hailes, Gloucestershire (Stalley 1987, 215; Park 1986, 204, pl. 86). This scene is on the south wall of the nave opposite a figure of St Christopher with the Christ child on his shoulders. The huntsman is sounding a horn and following a group of three lively hounds, in this case in pursuit not of a stag but of a hare (which can be seen to the left cowering under a bush). The iconography of stag hunting will be addressed further in the discussion of the Phase Two paintings below.

In Clare Island Abbey on the chancel's south wall, just above the door leading to the stairs, is a representation of a mounted horseman galloping eastwards (Pl. II). The small horse has a saddlecloth (of the same colour as the horseman's coat) extending over his croup and secured by two black girths, a crupper and a breast-girth. The snaffle bridle is also black. The strange curved black line in front of the horse's face does not

Pl. II   Phase One, south wall: horseman.

correspond to a chanfron, and consequently its role is unclear.

The rider is represented as a Gaelic horseman and is shown wearing a hauberk of overlapping squares, painted black with details incised and sometimes outlined in red. A fringe showing at its lower edge is probably that of the garment worn underneath (an *aketon*?). The rider wears a simple conical red helmet (basinet) without a nosepiece (Pl. III) and holds a long spear in his right hand and the reins in his left. A curious feature of the Clare Island horseman is his mail: such heavy armour would normally necessitate a high-backed saddle and a much larger horse. The pony, however, has a simple saddlecloth and, in true form for a Gaelic horseman, is without saddle or stirrups. Another curiosity is that the face and indeed the legs of this Gaelic horseman are painted black. The black colour used here is believed to be the same black as the armour rather than a discoloured pigment (Oldenbourg, this volume, p. 52).

It is possible that black was used as underpainting for the flesh tones of this figure and that the surface application has since deteriorated. This technique has been noted elsewhere on wall paintings of the early fourteenth century (Park 1986, 203)

There is a very close affinity between the Clare Island scene and that of Art Mac Murrough, the Gaelic-Irish Leinster leader, as illustrated by Jean Creton (Webb 1824, pl. 4; Stalley, this volume, pl. VIII, p. 142).[1] As with the Clare Island example, Mac Murrough is shown riding a small pony with no stirrups and wearing a conical helmet, a long robe and a suit of mail. The attire of the Gaelic party is clearly in contrast to that of the troops of Richard II (Mathew 1968, 209–10 and pl. 29, opposite p. 164). Mac Murrough's feet are bare. The figure on the horse at Clare Island may not be wearing shoes either. His leg is painted black, which may represent hose or, as suggested above, may be the underpainting for now deteriorated flesh tones.

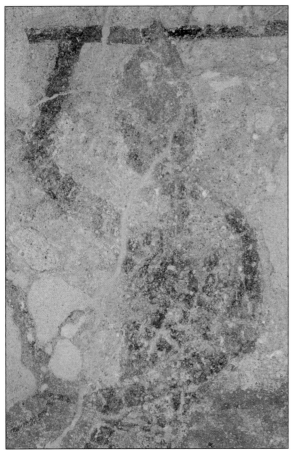

Pl. III Phase One, south wall: horseman—detail of upper body.

A tomb panel with a series of gallowglasses (professional Irish soldiers, often of Scottish descent) was inserted into the tomb of Felim O'Conor at Roscommon Abbey. The conical helmets and coats of mail worn by these figures (which are currently dated to the late fifteenth century (Hunt 1974, 216, Catalogue entry 213)) are very similar to those worn by the horseman at Clare Island, particularly in terms of the fringe and hem of their *aketons*.

Representations of figures mounted on horses are found elsewhere in Irish medieval art. A carved example on a stone at the church at Annagh, Co. Kerry, is somewhat similar, though the figure appears to be carrying a sword (Stalley, this volume, pl. IX, p. 143). Mounted knights are a common occurrence on Anglo-Norman seals (Stalley, this volume, p. 141), and there are small mounted soldiers with swords on the late medieval metalwork shrine called the Domhnach Airgid, which have been interpreted as representing the lordly Gaelic patrons the Maguires (Hourihane 2003, 136). A mounted huntsman features as part of the stag hunt depicted on the stone overmantle from Ardamullivan Castle (Pl. IV).[2] Again the horse or pony is small and low to the ground and has a saddle blanket or possibly a saddle. The dress details of the horseman are far from clear, but he has a conspicuous spur on his boot. He wields a long spear that has just reached its target: the injured stag is clearly in pain. The accompanying hound is in mid-flight and positioned just above the stag's head. Also, despite its considerably earlier date, it is hard to ignore the comparable range of scenes represented on the cross base from Oldcourt, Co. Wicklow (Ó hÉailidhe 1987).

Another striking similarity between the two scenes is that both figures hold a spear directly above their heads as if engaging in battle. In the case of Mac Murrough, he is not in battle but instead on his way to parley with the Duke of Gloucester. The mode of depiction here may have been intended to emphasise his military prowess. The similarity between Mac Murrough and the Clare Island horseman may be an indication that the latter was someone of rank, possibly the O'Malley lord.

There are considerable losses in the Phase One painting, and it is therefore difficult to establish

Pl. IV   Ardamullivan Castle, Co. Galway—fire mantle with a carved stag-hunt set in between vine leaves. A horseman holding a long spear chases a wounded stag with assailing hound in mid-flight (DOEHLG).

the relationship between the horseman on the south wall and the hunting scene on the north wall and, indeed, the nature of their relationship with the tomb niche. However, the tomb niche, with its elaborate canopy, is the central focus of the chancel, and the Phase One paintings were clearly intended to further adorn this chancel space and embellish the canopy, no doubt in accordance with the instructions of the benefactor. What is less clear is the message the benefactor wanted to convey through the painting and, indeed, the identity of this benefactor.

The salvation of the soul was of particular concern during the medieval period, and many religious houses were founded to house the tomb of the founder and his or her descendants (Leigh Fry 1999, 156). Tombs of founders or benefactors with associated wall paintings are not an uncommon feature of Cistercian abbeys. In Ireland they are found at Abbeyknockmoy, Jerpoint Abbey and Corcomroe Abbey.

The concept of visual commemoration is not unusual in medieval art. Indeed, animals are represented alongside patrons on funerary monuments. The loyal hound at the feet of its owner is a frequent occurrence on such effigies. At Clare Island the tomb is itself relatively plain, with no carved effigy or inscription that might help identify the patron. Is it possible that the horseman depicted on the south wall is, in fact, a representation of the patron and that he is included here as part of the hunting scene? Stalley (this volume, p. 143) has pointed out some parallels with sculpture in the Western Isles of Scotland. A stag hunt is included among a number of carved panels on the sixteenth-century tomb of Alexander MacLeod at Rodel in Scotland. MacLeod is depicted in a hauberk of mail, with a long *aketon* and a basinet. Two huntsmen attend him with four hounds in pursuit of three deer. Steer and Bannerman (1977, 186) suggested that while the costume represented was more in keeping with warfare than hunting, this attire was adopted to emphasise the status of the patron. This argument could apply also to the mode of representation of the horseman as Gaelic lord at Clare Island.

Phase One painting—conclusion

As part of the Phase One decoration, colour was applied to embellish the elaborate canopy and to decorate the adjoining and surrounding plaster. This decoration included the simple accentuation and simulation of architectural features with colour as well as a number of figurative representations. The canopy feature was clearly vital to the design of the Phase One painting; in fact it was significant enough for the artist to repeat its design in paint to the right of the stone version. The hound pursues the stag through the spaces between both the painted and stone versions of the canopy. The images of the Phase One wall paintings are in good agreement with the role of the chancel as a place of lordly burial as reflected by the depiction of the activities of hunting and the dress of the horseman as a Gaelic lord.

Phase Two wall painting

At some stage the Phase One painting was replaced by paintings on a much smaller scale but with a wider-ranging palette and a large repertoire of complex imagery. It is difficult to establish the order in which the vault painting was intended to be read. Equally, it is a puzzle to understand why, in a church context, scenes of an apparently secular nature are to be found mixed with other images with obvious religious meanings. The greatest survival of the Phase Two painting is found today on the vault. However, as can be seen from the more fragmentary remains that extend down the north and south walls and also on the east wall, this is part of what was originally a more comprehensive decorative scheme. Moreover, as suggested by Oldenbourg (this volume, p. 51), the colour scheme on the tomb canopy was still considered to be relevant, being retained and incorporated into the Phase Two design.

The vault is covered with an extensive range of images set out in the spaces between 'vault ribs', which were painted to simulate stone (Fig. 1, Pl. V). These painted ribs with corbel terminals (which bear a strong resemblance to the carved versions in the mother house at Abbeyknockmoy, Co. Galway) form a framework for the design on the vault.

Essentially the painted ribs (in much the same way as architectural ribs) have no bearing on the reading of the paintings except that they impose certain spatial constraints on the scenes in the interspaces. Many of the scenes extend either side

Pl. V   Phase Two: overall view of the vault painting. See also fold-out Illustrations 1 and 2.

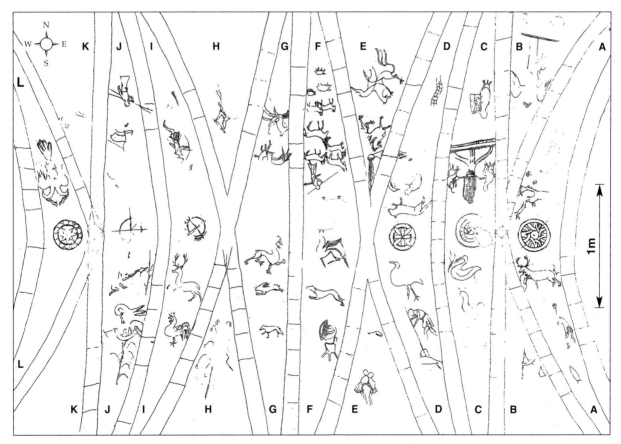

Fig. 1   Overall drawing of vault paintings (Christoph Oldenbourg). See also fold-out Illustrations 1 and 2.

Pl. VI Phase Two: south side (Section G)—the incised and painted details of the hare are at variance, indicating a slight change of mind on the part of the artist when he came to paint the image.

of the 'architectural ribs' across two or more sections. The design layout for the vault paintings is complex and appears to have been very well planned, and there are indications of at least two hands at work. As a first step a sketch of the proposed design was mapped out in the still wet plaster, much like a preliminary drawing. Most of the images were freely mapped out with incised lines, but the designs were generally corrected and improved as the work of painting proceeded. In some instances, however, the initial incised sketch is doggedly adhered to, suggesting a different hand, perhaps that of an apprentice. An example of this is the yellow hare in Section G (Pl. VI) on the south side of the vault and the herdsman in Section F on the north side of the vault (Pl. XXII). In other cases the paint is applied so freely that it underlies, abuts or extends on to the area of the painted ribs. This is seen, for example, on the north side where emissions/flames extend eastwards from the mouth of the serpent in Section D and where the spears of the figure in Section J overlap the rib to the west (see Morton and Oldenbourg, this volume, pl. XXXVI, p. 81). While this overlayering of paint cannot be seen from the ground level it was observed while working at close quarters during conservation.

It is difficult to understand why a particular scene is placed next to another within the architectural framework. If there is a connection it is seldom obvious. In what order were the scenes to be read? The line of the apex, punctuated at the east and west ends by decorative bosses, provides a natural division between the side viewed when facing north and that viewed when facing south. The two sides are, in effect, placed back to back. It is noteworthy that the preponderance of figures on the south side move from west to east, unlike the north side, where most figures move from east to west.

Westropp considered that the scenes were a heterogeneous grouping of images (Fig. 1, Pl. V). Certainly, at first glance this appears to be the case. There are, however, some composite scenes that span a number of sections, while other apparently disparate scenes may be shown to have links with those in adjoining bays and to contribute to an overall iconography. In consideration of the likely subject matter Westropp pointed to the scenes of clear or possible religious significance: 'The one unmistakably religious subject on the ceiling is the angel with the scales; but the wolf may symbolise Satan, the stag ("desiring the waterbrooks") the fervent believer, and the cock the need of vigilance, such symbolic designs being widespread' (Westropp 1911, 34). He also made mention of the symbolic pelican and a phoenix. The latter is probably the double-headed eagle that was fully uncovered during recent conservation.

Equally, there are a number of elements or motifs on the vault painting for which the religious symbolism is not obvious or apparent today. Perhaps the overall iconography and its significance, as conceived during the medieval period, will never again be fully understood. A clue to the meaning may be gleaned by examination of the disparate scenes, first taken individually and then considered collectively to establish a context. Analysis suggests the presence of a coherent narrative. In approaching the overall iconography, it may be as well to take a lead from the iconography that has known significance, such as St Michael weighing the souls, the pelican, the double-headed eagle and the cock.

Roe demonstrated that a strong devotion to St Michael persisted over a good number of centuries in Ireland, as elsewhere, as manifest in historical sources, in poetry and in all art forms of the period (Roe 1975, 251–64). In addition to scenes of St Michael contending with the dragon there are many examples of St Michael weighing the souls in Irish art. Biblical references to the archangel can be found in Daniel 5:27; 10:13 and

12:1 and in Revelation 12:7. Roe pointed to the fact that the occurrence and recurrence of St Michael in Irish art was visual evidence of 'the unchanging nature of man's anxieties and concern with the terrors of the world to come, and their continuing trust in the Archangel's care and protection' (Roe 1975, 255). These images reflect man's sense of personal unworthiness as well as his vision of Judgement Day and its dangers and his continuing hopes for personal salvation through St Michael's protection (Roe 1975, 256–8).

Representations of St Michael with the scales convey the message that on the Day of Judgement one's sins and good deeds are weighed against each other. In addition to the Clare Island St Michael weighing the souls (Pl. VII), two further examples are known from Irish wall paintings. A now almost completely destroyed figure in the vicinity of the O'Kelly tomb (where it is also associated with the Crucifixion) at Abbeyknockmoy, Co. Galway, was thought to be a figure with scales (Crawford 1919, 25). A more definite example is found on the wall paintings at Ardamullivan Castle, Co. Galway

(Morton 2001, 107). Carved stone examples of this subject are found on the west door of Clontuskert Priory, Co. Galway (Pl. VIII), and on a tomb side at Cashel Cathedral, Co. Tipperary.

Interestingly, Irish examples of the weighing of the souls, of all dates, whether carved or painted, do not appear to include the Virgin, who intercedes on the part of the good souls in so many English and Continental wall paintings. The Devil, in the form of a minor demon, is generally included in the weighing of the souls in England. At Clare Island the archangel is located in the darkest part of the vault at the west end of the north wall. Malgorzata D'aughton (pers. comm.) suggested that this positioning may have been intentional and may have followed convention as indicated by Bede in the *Lives of the Abbots*.[3] Today St Michael remains concealed by grey calcification and we still have to rely on Westropp's watercolour when considering the details (Pl. VII; see also pl. XVII in Morton and Oldenbourg, this volume, p. 72). This area is to be the subject of further tests, and it is anticipated

Pl. VII   Westropp's colour drawing of St Michael weighing the souls (RIA).

Pl. VIII   Clontuskert Priory, Co. Galway, St Michael weighing the souls.

that the image can be retrieved. Even now there are slight indications of incised lines, interpreted by some as a demon with a trident possibly swinging from the scales in order to weigh it down. Here the devil is red, while at Ardamullivan Castle and in the Book of Kells, the demon is generally painted black, employing the usual colour symbolism for evil.

Hourihane (2003, 100) noted that in medieval art great use was made of the nature and the behaviour of animals for didactic purposes. Three birds with religious symbolism—the pelican, the cock and double-headed eagle—are represented on the vault in Clare Island Abbey. A pelican is found in Section J on the south side: the pelican, which was thought to feed its young on its own blood, was commonly used to represent the Passion and Resurrection of Jesus Christ (Hourihane 1984, 70–1). Hourihane (2003, 101) has shown that the symbol of the pelican is very popular in medieval art, with no less than fifteen examples known in Ireland. At Clare Island the pelican, located close to the eastern end of the south wall, feeds its young in a nest that is shaped like a boat and perched on a tree or bush. This is the only known painted example of a pelican in Ireland. A pelican on the metal Ballymacasey cross from Co. Kerry, dated 1479, appears with the Crucifixion, as well as another bird, a fox and a hare (Hourihane 2003, 103–6). Pelicans are also to be found on the baptismal font from Dunsany, Co. Meath, (Roe 1968, 17) and the west door at Clontuskert Priory, Co. Galway (Pl. IX). Further carvings of pelicans can be seen on the carved screen at Kilcooly Abbey, the doorway at Lorrha Priory and two carvings from Holycross Abbey, all in County Tipperary (Hourihane 2003, 106–10).

The crowing cock, a symbol of Christ's Passion

Pl. X   Phase Two: north side (Section L)— double-headed eagle.

and the Day of Judgement, is represented twice on the south side of the vault at Clare Island, once in Section C and again in Section I (see Morton and Oldenbourg, this volume, pls XLI, XLII, p. 84; pls LX, LXI, fig. 26, p.#90).

The double-headed eagle, seen in Section L on the north side of the vault (Pl. X), reputedly symbolised the eastern and western divisions of the Holy Roman Empire: the double-headed eagle only appeared on the arms of the Holy Roman Empire in the early fifteenth century (Kirkpatrick 1992, 336). There is a reference to the double-headed eagle about the altar of God in St Bernard's Apologia (James 1951, 141). Westropp (1911, 34) described a bird on the Clare Island wall painting that he thought might possibly be a phoenix. He did not produce an illustration of the bird but did comment on the appalling condition of the plaster at the eastern end of the vault (Westropp 1911, 35)—the exact area where the double-headed eagle was uncovered during conservation. It is suggested here that the bird examined by Westropp is the double-headed eagle. The eagle, because of its powerful heavenward flight and gaze fixed on the sun,

Pl. IX   Clontuskert Priory, Co. Galway, Pelican.

Pl. XI   Abbeyknockmoy grave marker slab— double-headed eagle.

became a symbol of the Ascension in Christian art (Hulme 1975, 184). In the context of the other images on the vault, this symbolism would seem appropriate in such a position, over the east wall and altar. As Stalley has pointed out (this volume, p. 145), double-headed eagles are found on line-impressed tiles from a number of Irish sites. These sites all have an eastern distribution and include the three Cistercian sites of St Mary's, Mellifont Abbey and Dunbrody Abbey (Eames and Fanning

1988, 87). Two Galway families, the Joyces and the Brownes, used a double-headed eagle in their coats of arms, and this motif, referring to the Brownes, can be seen among others in a coat of arms carved on a late-seventeenth-century wall plaque of the French family at Abbeyknockmoy, Co. Galway (Fitzgerald 1895–7) (Pl. XI). It is possible that the double-headed eagle at Clare Island had an heraldic significance.

Hunting scenes, which at first glance might appear to be secular subject matter, are a common feature in medieval art. Hunting with hounds and hawks was regarded as an aristocratic sport in the middle ages (Sekules 2001, 175), and it is found represented in all media from manuscripts to pavement tiles, from carving to metalwork and from tapestry to wall painting. A tiny panel on the Stowe Missal shrine shows a hound attacking a deer, and, as mentioned before, hunting scenes occur in both phases of wall paintings at Clare Island. The interesting feature of the Clare Island paintings is that on the Phase Two painting there are several hunting episodes depicted. A number of different modes of representing hunting, in particular stag hunting, are employed. This raises the issue of whether different episodes of the hunt could have had different meanings. Two further examples of stag hunts in Irish wall paintings appear in Holycross Abbey, Co. Tipperary (Pls XII and XIII), and Urlan More Castle, Co. Clare. In the Clare Island paintings, on the south side of the

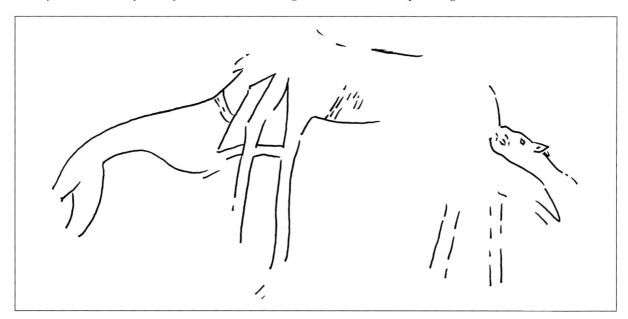

Fig. 2   Hunting scene at Urlan More Castle, Co. Clare. The drawing was made from a photographic slide taken in 1997.

Pl. XII Holycross Abbey, Co. Tipperary, west wall of north transept, part of a hunting scene.

Pl. XIII Holycross Abbey, Co. Tipperary, hunting scene—detail of huntsman and hound on slip-leash.

vault, there are two scenes where stags are pursued or attacked by hounds. These two scenes differ. That in Section A shows a stag being attacked by hounds at its neck and hindquarters (see Morton and Oldenbourg, this volume, pl. XXXVIII, p. 82) and is not dissimilar to that at Urlan More Castle (Gleeson 1936, 193). Urlan More Castle collapsed in February 1999, reducing the wall paintings to rubble. Details of the hunting scene can be gleaned from Gleeson's illustration (1936, 193) and close examination of more recent photographs (from 1997) of the wall paintings, which were then masked by microbiology and recent graffiti (Fig. 2). A standing stag, painted in yellow–orange and outlined in black, is attacked at the front by a black hound with a white collar and at the hindquarters by a white hound. The hound at the rear has its ears cocked, its mouth open and its teeth bared.

The second stag hunting scene on the vault at Clare Island is in Sections I and J (Pl. XIV). Here, as at Holycross (Crawford 1915, 151), the stag is

Pl. XIV Phase Two: Sections I and J—huntsman with a hound on a slip-leash and the hunted stag in the adjoining section.

separated from its hunters by a tree. Also in common with Holycross, the hound strains at the leash, but while at Holycross the slipper is standing (Pls XII, XIII), blowing a horn, and is followed by two archers, in the Clare Island example the slipper is crouching and alone. The details of the huntsman's dress at Clare Island Abbey are too small in scale and too deteriorated in condition to afford any specific description. It is difficult to know where one scene ends and the next begins. Is the scene with the hound in pursuit of a hare in Sections F and G (Pl. XV) part of the stag hunt as described above or is this yet another independent hunting scene?

It should be stressed that the stag hunt as an aristocratic pursuit, if considered in isolation, might not be out of place in Urlan More Castle. However, a clearly religious subject, the Madonna and child recorded by Wallace (1936–41, 38–9), as well as an inscribed design, possibly of a boat, formed part of the wall painting decoration there, and, as at Clare Island, we find a puzzling mixture of religious and apparently secular imagery.

Given the mixture of subjects portrayed at Clare Island, where some scenes and images are clearly of religious significance, one must at least consider the possibility that scenes of apparently non-religious meaning may originally have had religious symbolism. In addition to these scenes being indicative of the patron's high status they may well have had other layers of meaning. The separation of the religious and secular was not as marked during the Middle Ages as it is now. Certainly some castles had a room that functioned as a chapel, as at Barryscourt Castle, Co. Cork. Ironically, the wall paintings at Barryscourt do not

Pl. XV   Phase Two: south side—quarter view (Sections A–G) of vault with a range of scenes including a stag attacked at the neck and hindquarters by hounds; fishermen; a harpist; and a hound in pursuit of a hare.

include images of explicit religious content. In contrast, the painted chamber at Ardamullivan Castle has no features that would point to its use as a chapel and yet the wall paintings are exclusively and obviously religious in their subject matter.

The stag, an emblem of solitude and purity (Hulme 1975, 167) became in the Middle Ages a symbol of the 'soul thirsting for God'. Park (1986, 204) interpreted the stag hunt in the Gorleston Psalter as the stag (symbolising the Christian soul) being attacked by the 'ungodly'. The notion of good and evil, as suggested, is in good agreement with the symbolism of other images on the vault at Clare Island. The south side of the vault also shows what appears to be a cat climbing a tree, probably after a bird, and the hound in pursuit of a hare, as mentioned above (Pl. XV). A similar scene of hound and hare is found on the baptismal font from Dunsany, Co Meath (Roe 1968, 17), which also shows a pelican feeding her young and the crucifixion, as well as instruments of the Passion and a number of saints.

On the north side of the vault a hound chases

Pl. XVI   Phase Two: north side (section D)—a prowling lynx or wildcat.

Pl. XVII   Westropp's illustration of water stoup at Clare Island Abbey (RIA).

three hares; below is a possible falconry scene (in Section A). Section D contains a prowling incised lynx or wild cat (Pl. XVI): its prey in Section F is now lost. Below this a cat stalks a bird. Animals are also featured on the carved water stoup in the nave of the Abbey (Pl. XVII).

On the south side of the vault there are several fishing scenes with figures using tridents or poles (Pl. XV). Salmon-spearing on larger rivers and estuaries was a common method of fishing as late as the mid-nineteenth century, though it was outlawed from 1716 (Went 1952, 109). The trident was one of the most common forms of spearhead used (Went 1952, 125). This form of fishing would have been practised within the O'Malley lordship (though not on the Clare Island because of the lack of rivers there) and could have been an important source of income for the O'Malleys. It is difficult to establish a religious context for the inclusion of three fishermen on the vault paintings. The apostles were, of course, fishermen. Fish are used as an emblem of Christ and secondarily applied as a symbol of believers, based on the idea that Christ draws fish out of the 'waters of sin' (Hulme 1975, 197–8).

Pl. XVIII Phase Two: south side (Sections F and G)—affronted griffin and dragon.

Pl. XIX Clontuskert Priory, Co. Galway, affronted griffin and lion.

Other animals include the two fabulous beasts that face each other (Pl. XVIII). This scene bears a strong resemblance to a panel of an affronted griffin and lion from the west door at Clontuskert Priory, Co. Galway (Pl. XIX). A similar scene is found on a die-cast copper alloy object found near Durrow, Co. Offaly, now in the National Museum (Ó Floinn 2001, 307, pl. 14). Here the affronted lion and griffin are each set in their own panel and flanked by one panel with a stag and another with a hare.

There are, in addition to those mentioned above, many more animals painted on the vault at Clare Island. Hulme (1975) describes the symbolism of a whole range of animals. The lynx can denote suspicious vigilance; the goat, wolf and/or fox are associated with lust, cruelty and fraud. The ox suggests the sacrifice of Christ, the prophets, saints or apostles or even the sacrifice of a man who labours for the good of others. The horse can be an emblem of courage and generosity, while the pig, when used as a symbol of evil, can represent Satan. While Hulme described the swan as a symbol of hypocrisy, Hayden (1969, 11), in discussing the swan carving on the misericords of St Mary's Cathedral, Limerick, described it as a symbol of the martyrs, because the swan is said to sing with its dying breath. Each animal therefore could have had its own symbolism, which may have had a bearing on the decision to include them in the paintings.

It is hardly coincidental that, among other hunting scenes and scenes with animals, three stag hunts were included in the Clare Island imagery. Hence, it is important to view these images not as isolated motifs but as integral elements, albeit in this case with subliminal meaning.

Returning to the figure of St Michael, it is worth considering this in relation to the images in the neighbouring sections (Pl. XX). Immediately to the east are musicians—two figures playing and working an organ above a lyre player. Both musicians are facing east, and it has been suggested that in the context of the other images on the vault these may represent heavenly music for the good souls on their journey after judgement (Buckley, this volume, p. 127). Beyond the musicians on the north side is a serpent spitting flames or water across the painted rib into the next section. Serpents can have different

Pl. XX   Phase Two: north side (Sections C–G)—a range of scenes including musicians playing an organ and a lyre, a serpent, dragon, body-sharing beast and part of a 'cattle-raid'.

symbolisms—both good and evil—depending on the context (Kirkpatrick 1992, 928). Dragons on the other hand represent Satan—'the great dragon'. These fabulous winged beasts with a serpent's tail are as often called serpents as dragons (Kirkpatrick 1992, 318–19). In respect of the dragons in Section E, it must be remembered that St Michael is one of the saints particularly associated with dragon-slaying.

Above this dragon, at the top of Section E, what appear to be flames rise from a hand or gauntlet (Pl. XX). When he considered the weighing of the souls and the mode of depicting their ascension to heaven, Hulme (1975, 102) mentioned that the 'Torch of Life', though originally a pagan idea, occurs occasionally in Christian art. Perhaps this could be the significance of the hand with flames rising from it at Clare Island. In this context, the 'Tree of Life' was a favourite motif in the representation of the eternal struggle between good and evil, and one that dominates medieval art. It represents a cycle of growth—the tree blossoms, grows and loses

leaves, lies dormant and then regrows the following spring (Rouse 1972, 40). The tree in Section H on the south side of the vault may therefore be a reference to the 'Tree of Life', and this would add further weight to the idea of the hand with flames representing the 'Torch of Life'. The fruit tree could also have been seen as a reference to the O'Malley territory of Umhall. This sounds the same as the Irish word for an apple (*ubhall*) and was on at least one occasion translated into Latin as *pomum* (fruit), with the chief being referred to in the State Papers in 1515 as O'Malley *de pomo* (Hardiman 1846, n. 117).

The interchangeable use of the words 'serpent' and 'dragon', as seen in Revelation chapters 12 and 13, can be confusing and no doubt led to confusion in artists' representations. A serpent with water issuing from its mouth is referred to in 12:15—'Then from his mouth the serpent spewed water like a river, to overcome the woman [and in turn Christ, whom she had given birth to] and sweep her away with the torrent'. If this reference can be taken as the source for the serpent in Section D, perhaps then the serpent is aiming at the 'Torch of Life' in an attempt to quench the flames.

Elsewhere in Revelation are numerous passages that may have influenced the choice of subject matter depicted in the area immediately around St Michael. These include references to the dragon going to make war against those who obey God's commandments (12:17); to the dragon empowering a beast (13:2); and to those who refused to worship the image of the beast, who were to be killed (13:15). Music also features, with mention of harpists (Pl. XXI) playing and singing to those who had been redeemed from earth (14:2–3).[4]

The north wall also includes a composite scene, in Sections F to J (Pls XX and XXII). In Section F a herdsman and his animals move to the west. Among the animals driven by the herdsman are cattle, goats or sheep and pigs. An arrow, presumably fired by the archer to the east, has shot the herdsman. This detail helps to link the various sections of the scene. Indeed, the horse in Section G has also been shot with an arrow. Some years ago Kieran O'Conor (pers. comm.) suggested that the composite scene might be a cattle raid. Cattle-raiding themes pervade almost every branch of Irish literature, and cattle raiding

Pl. XXI   Phase Two: south side (Section F)—harpist with a running hound above it.

Pl. XXII   Phase Two: north side (Sections F–J)—vault with a composite scene of a 'cattle-raid'.

was regarded as a particularly aristocratic occupation right up to the seventeenth century (Lucas 1989, 123–97).

No unambiguous representations of cattle raids have yet been identified in Irish high art. Animal drives or herding scenes of a much earlier date do occur, however, on a number of carved stone crosses. An assortment of individual animals is included in these depictions, many of which can be seen on the vault at Clare Island— stags, sheep, geese (or perhaps swans), hares, dogs, cattle, horses, and a seal (?), as well as affronted animals and animals in combat. The herdsmen often carry spears and occasionally carry shields. In one instance, on the base of a cross from Oldcourt, Co. Wicklow, such images are juxtaposed with St Michael weighing the souls and with affronted animals, as well as other scenes (Ó hÉailidhe 1987, 98–110). Therefore the Clare Island cattle raid, if such it is, provides an interesting hint at the continuity of certain types of imagery even where the symbolism may have changed.

The eternal struggles between good and evil were variously portrayed in medieval art, and armed figures, sometimes on horseback, are a common occurrence in early examples (Rouse 1972, 39–40). It could be speculated that the warriors in the Clare Island Abbey paintings, in such close proximity to St Michael, provide an extension and consolidation of the theme of good and evil that prevails in these paintings. This sort of portrayal would underline the concept that fighting is sometimes necessary to combat evil. Given the importance of cattle raiding to Irish medieval Gaelic society, it should be no wonder that such a practice (which might perhaps be the artist's only reference for a combat scene) could be the subject of a wall painting.

At the beginning of the conservation programme it was discovered that overlying limewashes occurred solely in Sections I and J (on the north side of the vault)—that is directly over the 'cattle raid' (Oldenbourg, this volume, p. 57). The extent of the limewash was limited, and it is interesting to speculate on its purpose.

Two possibilities present themselves. The first, that the limewash was applied in the hope of saving and protecting an area of degraded plaster. However, there are no records of such an intervention, and if it was carried out this must have occurred long before the time of Westropp's visits for the first survey. The second possibility is that at some point in the history of the Abbey the images of fighting warriors were covered up, perhaps because they were considered offensive and not in keeping with a place of worship.

A few other images, including the wrestlers (on the south side in Section E; Pl. XXIII) and the four body-sharing beasts (on the north side, also in Section E) have yet to be considered. Wrestlers can symbolise the struggle with evil. Human wrestlers placed in the vicinity of the cock in Section C may amplify the sense of struggle between good and evil. The cock in Christianity became the symbol of Christ's victory over the power of darkness (M. D'aughton, pers. comm.).

The Clare Island wrestlers bring to mind a number of illustrations of wrestlers in the sketchbooks of Villard de Honnecourt and in other late Gothic examples. These illustrations demonstrate the same concern for balance. The figures are shown in a stylised fight as equals. Wrestlers are not uncommon in Irish medieval art,

with early examples on the Kells Market Cross, Co. Meath, and among the Romanesque carvings at Kilteel, Co. Kildare. A similar image is found on a wall painting in the Castello di Sabbionara, Italy (Provincia Autonoma di Trento 2002, 298, pl. 6), where the wrestlers are represented in the lower register of the painting. The colour treatment of yellow and red is similar to that used on the Clare Island painting except that the hose of the figure in a red tunic are yellow while the hose of the figure in a yellow tunic are red. Another similar example of wrestlers is that of Jacob wrestling with the angel as represented on the Brescia casket (Grabar 1968, pl. 336).

Villard de Honnecourt drew motifs similar to the four body-sharing animals (Pl. XXIV). There are numerous Irish examples of humans and animals, real and fabulous, that are interlaced. The beasts here, however, share their forequarters with their nearest neighbour to one side and their hindquarters with the nearest neighbour on the other side in a manner reminiscent of the 'Tinners' Rabbits' (Anna Hulbert, pers. comm.). The 'Tinners' Rabbits', which are found on the late fifteenth-century wooden roof bosses in a number of Devon churches, share ears with their neighbours. A possible parallel for the Clare Island motif from an Irish context is found on the

Pl. XXIII   Phase Two: south side (Section E)—wrestlers.

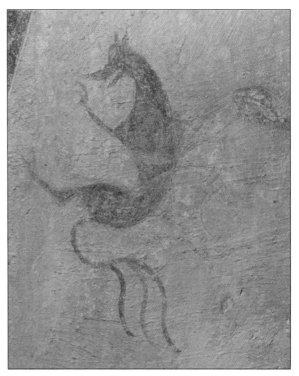

Pl. XXIV   Phase Two: north side (Section E)—four body-sharing beasts.

leather satchel of the Book of Armagh (Hourihane 2003, 149, pl. 180). Here three rather than four beasts share body parts: the front leg and foot of one beast is the abdomen and back leg of another.

Phase Two Painting—conclusion

The Phase Two scheme, including the ribs and motifs for the interspaces, were all scribed into the still damp plaster. Slight alterations were accommodated either by incorporating changes when the plaster was wet or a little later when the colour was being applied (Pl. VI). There is, however, no evidence for a complete change of mind in the representation of even a single motif. The overall iconographic and design concept was well planned and most probably was worked through, tried and tested by the artists and approved by the patron before the plaster was ever applied to the vault. It might even be suggested that such a comprehensive and complex design would have necessitated at least one preliminary drawing.

The exact order in which these images on the vault were to be viewed remains uncertain. Similarly we cannot be sure why religious imagery appears alongside apparently secular, bucolic (if not aristocratic) scenes of hunting and real or imaginary beasts on the vault of a church. Certainly the inclusion of motifs that are indicative of lordly status among religious images appears both intentional and uninhibited. The evidence from the Phase Two painting at Clare Island would suggest that the divide between the secular and religious components of a Christian's life was not as marked or distinct during the late medieval period as might be assumed today, and perhaps for this reason attempts to define the role of the wall painting and to classify the images as religious or non-religious have been frustrated.

Hourihane (2003, 12) noted the difficulty in interpreting the art of the late Gothic period due to the ever increasing number and complex range of motifs, which were drawn from a variety of sources. While the intended Clare Island design was clearly thought through, as evidenced by the relatively close adherence to the incised drawing during the subsequent application of paint, the compositional elements within the design are not clearly defined. This is further complicated when the extent of particular groupings of images is not indicated. As mentioned, it is not clear whether the vignette of the hound chasing the hare is part of the stag hunt in the adjacent sections or whether they stand as two distinct scenes. This apparent incoherence from vignette to vignette might paradoxically add weight to any suggestion that when the images were originally portrayed their arrangements and their significance were clearly understood by artist, patron and viewer alike.

I would argue that the grouping of images is far more homogeneous than has been suggested by Westropp (1911, 33–4) and others. From a religious perspective, the overall theme might be one of the Last Judgement and ultimately of salvation, with St Michael acting as Christ's judge on earth—weighing good and evil souls and determining their fate in the afterlife. Again, the crucifixion scene on the north wall (on the back wall of the tomb niche) shows Christ the Redeemer, whose death was the ultimate sacrifice for man. This overall concept has been confirmed and amplified by recurring references throughout the paintings. However, while the collective impression from these images is of the contrast and the conflict between good and evil, it still remains for the 'secret code' to be broken and for the deeper layers of meaning of some episodes within the decorative scheme to be understood.

Many of the images on the vault could have been taken directly from observation of daily and seasonal practices in the O'Malley lordship, from high-status activities such as hunting, cattle raiding and the making of music, to more lowly activities such as fishing. What points of reference were there for the fabulous and imaginary beasts such as the dragons, griffins, body-sharing beasts and lynxes? Sources probably included decoration on mobile artefacts, books of devotion and carvings or paintings observed by the travelling artist or monk. Given the similarity in the images of the wrestlers and, indeed, the body-sharing beast from Clare Island with the drawings of Villard de Honnecourt, it might not be unreasonable to suggest that the artist had seen or had in his possession a 'pattern book' of sorts. The pelican, which has no biblical authority, has a long pedigree of Christian symbolism (Hulme 1975, 182), but one need go no further than the Bible to account for much of the imagery at Clare Island.

While parallels for individual Clare Island motifs are not hard to find in late medieval art in Ireland, it is more difficult to identify a site with a comparable range of motifs. The cloister arcade at Jerpoint Abbey has a huge range of carvings and motifs (Rae 1966), but they do not relate closely to the images at Clare Island. Perhaps the west portal at Clontuskert Priory, Co. Galway (Pl. XXV), provides the closest parallel, although it has just three comparable scenes and some other similar motifs in common with the Clare Island paintings. This portal has an inscription that includes the date 1471 (Hourihane 2003, 72).[5] The three motifs in common with Clare Island are St Michael Weighing the Souls (with two little devils tipping the scales); the pelican in piety; and an affronted lion and griffin. In contrast to the lion and griffin at Clontuskert, the equivalent vignette from Clare Island depicts an affronted griffin and dragon. Another panel from Clontuskert in which a bird and beast as well as the devil are represented at the feet of St Catherine would not look out of place in Clare Island Abbey. The Clontuskert portal provides links with other sites that have images comparable to those at Clare Island as well. For example, the Clontuskert portal has a panel with two animals (yales)[6] with entwined necks similar to that at Holycross Abbey (exterior north transept) and a mermaid with mirror, and a pelican similar to those at Kilcooly Abbey.

While the Clare Island paintings show a familiarity with Christian symbolism and with a large repertoire of stock images seen elsewhere in the art of medieval Ireland and Europe, Stalley (1987, 216) has described the Clare Island paintings as particularly Gaelic, and, indeed, they do appear to have specific Irish reference. Some of the animals, in particular the hound in Section F on the south wall, are reminiscent of animals represented in early Irish illuminated manuscripts. However, the Clare Island wall paintings do have a European dimension, and while the specifics are Irish, the types of images and their symbolism fall within a well-established European tradition.

Dating the Clare Island wall paintings

Examination of the wall paintings and their relationship with each other and the structure of the Abbey provides evidence for two distinct construction phases of the chancel, which correlate with the two different wall painting phases.

Clearly the dating of the Phase Two paintings relies on the dating of the Phase One paintings for its *terminus post quem*. The dating of the Phase One painting is inextricably linked with the dating of the canopy feature in the north wall, where the plaster of the Phase One painting abuts the masonry canopy of the tomb niche (Pl. XXVI). This painted plaster provides proof that the tomb niche pre-dates both painting phases, but by how long is uncertain.

There are no inscriptions to date the elaborate canopied tomb (Pl. XXVI). Manning (this volume, pp 23–24) acknowledges that it is difficult to provide a tight date range due to an imperfect understanding of this type of monument. Traceried canopies are a fifteenth- and sixteenth-century feature in religious foundations, with a few examples that might be dated to the middle or early years of the 1400s. Clearly, the elaborate canopied tomb in the north wall is central to the first phase of decoration. It is likely to be the tomb of the patron who financed the building of the

Pl. XXV   Clontuskert Priory, Co. Galway, overall view of the west portal.

Pl. XXVI North Wall: tomb niche and elaborate canopy on the north wall of the chancel (DOEHLG).

chancel, but it could also have served as an Easter sepulchre (Duffy 1992, 32).

The discovery of the painted copy of the canopy further east on the north wall is very significant (see Morton and Oldenbourg, this volume, pl. I, p. 62; fig. 1, p. 63). Firstly, it copies the carved canopy in every detail, with the blocks of the tracery indicated by black joint lines on a red background. Secondly, an examination of the few remaining traces of paint on the carved canopy itself illustrates that the different faces of this 'prototype' were decorated with the same colours as the painted version. Thirdly, the dimensions of the carved and painted canopies are the same: they each take up exactly one third of the length of the north wall, where the tomb niche was centrally placed. It might be postulated, therefore, that a second painted version of the canopy was originally placed to the west of the carved canopy, thereby occupying the remaining third of the wall space. The paint remains in this area are largely concealed by the relatively complete survival of the overlying Phase Two plaster and therefore offer no support for this argument. It has been shown that the painted design for the north wall incorporated the carved canopy feature in an overall decorative scheme (Morton and Oldenbourg, this volume, pp 63–65), and it is clear that this layout was planned from the outset, with the canopied tomb centrally placed among the painted decoration. It is reasonable to speculate on this evidence that the tomb and the Phase One paintings are contemporary.

Lastly, and possibly most importantly, the painted version of the canopy appears to run behind and be partially blocked by a load-bearing pier carrying a relieving arch: this arch in turn helps to support the east end of the vault, which bore most of the Phase Two paintings. It may be argued that the pier post-dates the Phase One paintings on the north wall and therefore that the current vault was a later insertion.

Further building alterations in the chancel can be detected on the south wall. The horseman is located at the western end of the wall, above the door leading to the stairs. The plaster (Phase Two paintings) on the stone arch of the door below is proud of the Phase One surfaces and partially overlaps the horseman (Pl. II). The arch must therefore post-date the earlier scheme of paintings. The original entrance to the stairs was through the low door from the nave.

It is clear that several building alterations were carried out after the completion of the Phase One paintings and before, or more probably in association with, the creation of the Phase Two paintings. The Phase One paintings were housed in a relatively simple chancel, possibly with a flat wooden ceiling and with an elaborate tomb niche centrally placed in the north wall.[7] The chancel was later altered—an opening to a new room was made in the north wall, a door giving access to the stairs was added to the south wall and a stone vault was constructed (over wicker centering). The walls and vault were then plastered and decorated with a second scheme of wall paintings (Manning's Construction Phase 3 (this volume, p. 24)).

The technique of constructing a barrel vault over wicker centering has been well documented and illustrated (Leask 1977, 86–7). Details of the construction technique used at Clare Island were observed and documented during the conservation programme. As mentioned above, very little of the wicker centering survived, and it has been postulated that it might well have been removed to provide a better key for the subsequent plaster (Oldenbourg, this volume, pp 50–51). One of the few small fragments of wicker that were found was sampled for radiocarbon dating, which was carried out at the Research Laboratory for Archaeology and the History of Art, Oxford (OXA-6951), giving a calibration date of AD 1345±75, which was calculated with a 95% confidence level (Hedges *et al.* 1998, 237). Evidently, there is a 5% chance that the vault was constructed outside of this date range. In itself, therefore, the radiocarbon date is not specific enough but can serve only as a broad indicator.

The existence of two relatively comprehensive decorative schemes in this small church raises many questions: Who commissioned these paintings? What role do they fill or what story do they tell? Was their purpose educational, devotional or commemorative, for instance? What factors influenced or what event necessitated the replacement of the earlier painting? What period of time intervened between the completion of the Phase One paintings and the refurbishment of the chancel? What dates can be ascribed to the two different painting phases? None of these questions can be easily answered.

As with other ecclesiastical buildings, it is likely that one or more benefactors funded the foundation and subsequent alterations of the Abbey at Clare Island. Together with the construction of the chancel, provision was made for an elaborate tomb niche and the Phase One wall paintings. In this way the benefactors secured themselves and their descendants a final resting place and a memorial.

The Annals of Connacht of 1415 note a branch of the O'Malley family under the leadership of a Diarmait Ó Máille, who defeated the family of the previous lords in 1415 and established a new dynasty.[8] An account by R. Downing from about 1684 claims that Clare Island Abbey was built by 'Dermitius (Claudus) O Maly' and that he and his wife (Maeve O'Conor) and many of their descendants are buried there (Manning, this volume, p. 10). If the Diarmait Ó Máille and 'Dermitius (Claudus) O Maly' in the above references are one and the same person, and we assume that Downing's source was reliable, a building date for the chancel (Manning's

Construction Phase 2) of about 1420–30 could be proposed (Manning, this volume, pp 22–24).

Who then was the benefactor who provided for the chancel's alterations, including the insertion of the vault and the Phase Two wall paintings? In addition to securing a burial place, the patron was concerned with assuring the salvation of his soul, as the iconography of the Phase Two paintings suggests.

Manning's Construction Phase 3 and its associated Phase Two wall paintings represent a major alteration and elaboration of the existing chancel. In fact, the layout for the design of the Phase Two painting relied for its success on being applied to a curved surface of the vault rather than a flat ceiling. The false vault ribs here, so similar to the carved ribs at Abbeyknockmoy (an O'Conor foundation, with later O'Kelly patronage), reinforce the historically documented connection between the two places. Whether this connection indicates links between the monks, the patrons or the artists at both sites remains a matter of speculation.

Interestingly, the polychrome painting of the carved canopy was not covered over by the Phase Two paintings, although every other available wall surface was re-plastered and painted. This is significant, as it suggests that this polychromy on the wall tomb was still in good condition and, more importantly, was considered suitable to be incorporated in the later painting scheme. This further suggests that the time span between the Phase One and Phase Two paintings was not very great, as the polychrome painting on the carved elements would have begun to show signs of wear and tear and would most likely have been repainted had much time elapsed between the two phases.

The differences between the two painting phases have been discussed above. These phases have, however, a few details in common. Whatever the time lapse between the two phases may have been, hunting scenes are a feature of both and were clearly still in vogue and pertinent when the Phase Two paintings were made.

Costume can be an important aid to dating as people were generally depicted in contemporary dress to help recognition of rank, status or occupation (Rouse 1972, 6). There is just a single figure from the Phase One painting and a number of small-scale figures from the

Phase Two painting that can provide details of costume.

Dunleavy noted hoods with liripipes in the depiction of Art Mac Murrough and his aides, a dress detail introduced to England and Ireland around 1300 (1989, 35–7). The liripipe is not a feature of the costume of the horseman at Clare Island, perhaps suggesting the continued obstinacy of the Irish in matters of dress, especially in this area so far removed from the Pale (Pl. III). Despite this, the striking similarities between Art Mac Murrough and the Clare Island horseman referred to above cannot be ignored. If the attribution of the tomb in Dungiven Priory, Co. Derry, to Cooey na Gall O'Cahan and therefore its dating to the late fourteenth century (McNeill 2001, 348–52) is accepted this would provide a possible date for the carved panel of gallowglass figures both at this site and in Roscommon Friary. The similarity in the details in the coats of mail at Roscommon and at Clare Island has been mentioned above.

The figures in the Phase Two wall paintings are so small and in places so deteriorated that details of their attire are difficult to determine. Some figures wear knee-length skirts or tunics with hose, while others wear three-quarter-length garments (Pls XV, XXII and XXIII). In Section H on the north side, the figure in this part of the cattle raid is wearing a conical basinet of similar shape to that of the horseman in the Phase One paintings. Those figures on the vault clad in yellow garments may represent the Irish in their tunics of saffron, considered a distinctively Irish dye (Dunleavy 1989, 47).

A number of figures in the 'cattle raid' on the north side of the vault carry spears similar to that of the horseman in the Phase One paintings. Although the herdsman and one of the horses is shot with an arrow, only one figure with a bow survives on the vault today. The bow, held at arms length, reaches from the top of the head to just above knee height and is similar to those represented in the hunting scene at Holycross Abbey and in the Martyrdom of St Sebastian at Abbeyknockmoy. But, again, the details of the Clare Island bow are not easily discerned.

Precise dating remains elusive. A suggested date range from the first few decades of the fifteenth century for the Phase One painting and a mid-1400s date for the Phase Two paintings would not sit uncomfortably with the evidence presented. Certainly, these dates would link well with the general expansion of friaries and priories being built by Gaelic lords. This period saw major building, resulting in part from the new confidence of the Gaelic-Irish after the resurgence of the fourteenth century (Leask 1960, 1–9).

Conclusion

The Phase One wall paintings allude perhaps more to the patron's lordly position on earth rather than a concern about his afterlife. The inclusion of a painted representation of the canopy is significant and reinforces the importance of the centrally placed carved 'prototype' in the overall decoration of the chancel.

There is little in the iconography of the Phase One paintings that helps to pinpoint a date. The costume of the horseman cannot be securely dated. Because of similarities in some of the elements between the tomb canopy at Clare Island and two tomb canopies at Kilconnell, Co. Galway,[9] and a number of other monuments elsewhere, a date in the early to mid-fifteenth century could be proposed for the Phase One wall paintings.

The Phase One paintings were superseded by a new and extensive painting that covered the walls and vault of the chancel. It is argued that the layout of the vault painting is far more homogeneous than might be thought on first or even second glance. While some motifs have certain religious symbolism, others find reference in nature and contemporary activities. It is more difficult to explain or identify the more fabulous beasts. They too must be symbolic. These images seem to occupy a space somewhere between the secular and religious and between the real and imaginary worlds. It is also possible that some had heraldic significance.

The canopied tomb in the north wall could have also served as an Easter Sepulchre, and this emphasises the concern with death that appears to have persisted through both phases in the chancel. This concern is reflected by the continuity in the use of the tomb, by its embellishment and, in particular, by the message of the Phase Two paintings.

A theme of judgement and salvation is postulated for the Phase Two work. However, the interpretation of many of the individual vignettes is, and may always remain, speculative. The contrast between good and evil and the presence of Christ as man's redeemer are of paramount importance throughout. Such themes dominated the art of the later medieval period.

It is difficult to understand the exact context for all the component images and the role or purpose of the paintings, but the meaning must in some way relate to the patron. They may have been didactic or devotional. Equally, the painting may have acted as a means of promoting the salvation of the patron's soul. While comparable examples for some of the individual scenes can be found in other church wall paintings—hunting scenes at Holycross Abbey and Urlan More Castle and St Michael weighing the souls at Ardamullivan Castle—the Phase Two painting at Clare Island is so far unique in the range of its motifs and the manner of its depiction. The closest parallels for the different elements in the Phase Two painting are to be found in the stone carving on the west portal at Clontuskert Priory.

There are several indicators that point to a date of between the 1400s and 1450s for the Phase Two paintings. The radiocarbon date obtained from a sample of vault wicker acts as a broad indicator of date, and this tallies with the other evidence. As mentioned above, it is postulated that the polychromy on the tomb niche remained in good condition and was included as part of the Phase Two paintings, thereby suggesting a time frame of perhaps as little as a decade or two between the two decorative schemes. The reference to Diarmait Ó Máille is of interest, although it cannot be established with certainty whether it has any bearing on one or other of the painting phases. The subject matter of the Phase Two paintings themselves is not a very useful indicator of date. Unfortunately, very few parallels for these motifs come from securely dated contexts. The west portal at Clontuskert Priory, dated 1471, provides some parallels, but many of these motifs were used over a long time period.

It could be said that the work undertaken by Westropp for the first Clare Island survey and the recent conservation programme together provide the foundation for a clearer understanding of all the aspects of these remarkable paintings.

The first step, and the most fundamental, was that of identification and recording. The second step, that of conservation, has provided an opportunity to stabilise and uncover the extant remains, to examine and record them in more detail and to present them in such a way that every aspect can be studied closely and appreciated today and in the future. It might be argued that the conserved paintings raise more questions than they answer. They certainly provide ample material for academic research and debate and—equally importantly—for appreciation.

Acknowledgements
Firstly I would like to acknowledge the National Monuments Service for commissioning the wall painting conservation programme. The contribution of the conservation programme at Clare Island cannot be over emphasised. I would especially like to thank the conservation directors, Madeleine Katkov and Christoph Oldenbourg, for saving these wall paintings. This article could not have been attempted without their experience, skill and dedication to the project, and they have revealed for us all the fascinating characteristics of the paintings. I would further like to thank Christoph Oldenbourg for providing most of the photographs used in this paper and to thank the Royal Irish Academy for permission to reproduce Westropp's images in Pls VII and XVII.

I am indebted to a number of people who have been kind enough to read and comment on drafts of this text and alert me to bibliographic or visual references: Loughlin Kealy, School of Engineering and Architecture, UCD; Roger Stalley, Department of Art History, Trinity College; Malgorzata D'aughton; and my husband, Kieran O'Conor, Department of Archaeology, National University of Ireland, Galway. I would also like to thank Barbara Schick for drawing my attention to the Castello di Sabbionara wrestlers.

I would like to thank all the editors and external readers, including John Waddell, Department of Archaeology, NUI, Galway; and particularly Mr Conleth Manning of The National Monuments Service for all his helpful comments and suggestions and his infinite patience.

NOTES

1. See also the discussion by Stalley in the final section of this volume (pp 141–143).

2. This overmantle was discovered face down built into the wall walk of the castle during recent restoration work by the National Monuments Service. Carved vine leaves that feature at either end of the mantle appear to match those on the north wall window panels in the fifth storey hall as illustrated by Leask (1977, 96). This fifth storey is part of a later construction phase at Ardamullivan. The original tower was two-sectional and built up to three storeys.

3. When describing a set of paintings bought by Benedict Biscop between 679–680 for the church of St Peter's in Wearmouth, Bede explicitly stated that scenes from St John's vision of the Apocalypse were placed on the north wall. '... similitudes of the visions in the Revelation of the Blessed John for the ornament of the north wall in like manner, in order that all man which entered the church, even if they might not read, should either look (whatever way they turned) upon the gracious countenance of Christ and His saints, though it were in a picture; or might call to mind a more lively sense of the blessing of the Lord's incarnation, or having as it were before their eyes, the peril of the last judgement might remember more closely to examine themselves.' (King 1930, 405–7).

4. In 13:2 the serpent has reverted to a dragon 'The dragon gave the beast his power and his throne and great authority' and in 13:4 'Men worshiped the dragon because he had given authority to the beast and they also worshipped the beast...'.

5. Hourihane is mistaken in stating that this doorway has been reconstructed.

6. Yale: a heraldic antelope-like beast about the size of a horse with tusks, long horns and the tail of a lion or goat.

7. Manning's Construction Phase 2, the nave being Construction Phase 1 (this volume, pp 22–24).

8. In this reference Diarmait Ó Máille defeated Aed Ó Máille, king of Umall (who was killed during the raid): 'Diarmait O Maille was afterwards made king, and from this time the kingship and lordship departed from the seed of Aed.'

9. These tombs are undated but could be as early as the mid-fifteenth century, though Stalley's comments on the dating of this type of tomb generally should be noted (this volume, pp 139–141).

REFERENCES

Crawford, H.S. 1915 Mural paintings at Holycross Abbey. *Journal of the Royal Society of Antiquaries of Ireland* **45**, 149–51.

Crawford, H.S. 1919 Mural paintings and inscriptions at Knockmoy Abbey. *Journal of the Royal Society of Antiquaries of Ireland* **49**, 25–34.

Duffy, E. 1992 *The stripping of the altars: traditional religion in England c. 1400– c. 1580*. New Haven and London. Yale University Press.

Dunleavy, M. 1989 *Dress in Ireland*. London. Batsford Press.

Eames, E. and Fanning, T. 1988 *Irish medieval tiles*. Dublin. Royal Irish Academy.

Fitzgerald, W. 1895-7 Abbeyknockmoy, Co. Galway. *Journal of the Association for the Preservation of the Memorials of the Dead in Ireland* **3**, 277–8.

Gleeson, D.F. 1936 Drawing of a hunting scene, Urlan Castle, Co. Clare. *Journal of the Royal Society of Antiquaries of Ireland* **66**, 193.

Grabar, A. 1968 *Christian iconography: a study of its origins*. New Jersey. Princeton University Press.

Hardiman, J. 1846 *A chorographical description of West or h-Iar Connaught written in A.D. 1684 by Roderick OFlaherty, Esq*. Dublin. The Irish Archaeological Society.

Hayden, J.A. 1969 *Misericords in St Mary's Cathedral, Limerick*. Revised by Rev. M.J. Talbot, Dean of Limerick, 1969. Limerick Leader.

Hedges, R.E.M., Pettitt, P.B., Bronk Ramsey, C. and van Klinken, G.J. 1998 Radiocarbon dates from the Oxford AMS system archaeometry datelist 25, *Archaeometry* **40**, 227–39.

Hourihane, C. 1984 *The iconography of religious art in Ireland, 1250–1550*. Unpublished PhD thesis at the Courtauld Institute of Art, University of London.

Hourihane, C. 2003 *Gothic art in Ireland, 1169–1550*. London and New Haven. Yale University Press.

Hulme, E. 1975 *Symbolism in Christian art*. 2nd edn. Guildford. Biddles.

Hunt, J. 1974 *Irish medieval figure sculpture, 1200–1600*, 2 vols. Dublin and London. Irish University Press/Sotheby Parke Bernet.

James, M.R. 1951 *Pictor in carmine*. Archaeologia **94**, 141–66.

King, J.E. 1930 *Baedae Opera Historica*, Vol. II, 405–7. London and New York. Harvard University Press.

Kirkpatrick, B. (ed.) 1992 *Brewers concise dictionary of phase and fable*. Helicon Publishing Ltd.

Leask, H.G. 1960 *Irish churches and monastic buildings, Volume 3; medieval Gothic, the last phases*. Dundalk. Dundalgan Press.

Leask, H.G. 1977 *Irish castles*. Dundalk. Dundalgan Press.

Leigh Fry, S. 1999 *Burial in medieval Ireland 900–1500*. Dublin. Four Courts Press.

Lucas, A.T. 1989 *Cattle in ancient Ireland*. Kilkenny. Boethius Press.

Mathew, G. 1968 *The court of Richard II*. London. John Murray.

McNeill, T. 2001 The archaeology of Gaelic lordship

REFERENCES

east and west of the Foyle. In P.J. Duffy, D. Edwards and E. Fitzpatrick (eds), *Gaelic Ireland: land, lordship and settlement c. 1250–c. 1650*, 346–56. Dublin. Four Courts Press.

Morton, K. 2001 Medieval wall paintings at Ardamullivan. *Irish Arts Review Yearbook* **18**, 104–13.

Morton, K. 2004 Irish medieval wall painting. *Medieval Ireland, the Barryscourt Lectures*, 311–49. Kinsale. Gandon Editions.

Morton, K. forthcoming Wall painting report. In Miriam Clyne (ed.), *Excavations at Kells Priory, Co. Kilkenny*. Dublin. Stationery Office.

Ó Floinn, R. 2001 Goldsmiths' work in Ireland, 1200-1400. In C. Hourihane (ed.) *From Ireland coming: Irish art from the early Christian to the late Gothic period and its European context*, 289–312. New Jersey. Princeton University Press.

Ó hÉailidhe, P. 1987 The cross-base at Oldcourt, near Bray, Co. Wicklow. In E. Rynne (ed.), *Figures from the past: studies on figurative art in Christian Ireland, in honour of Helen M. Roe*, 98–110. Dun Laoghaire. Grendale Press and Royal Society of Antiquaries of Ireland.

Park, D. 1986 Cistercian wall painting and panel painting. In D. Park and C. Norton (eds), *Cistercian art and architecture in the British Isles*, 181–210. Cambridge. Cambridge University Press.

Provincia Autonoma di Trento 2002 *Catalogo a cura di Enrico Castelnuovo, Francesca de Gramatica: Il Gotico nelle Alpi 1350–1450*.

Rae, E.C. 1966 The sculpture of the cloister of Jerpoint Abbey. *Journal of the Royal Society of Antiquaries of Ireland* **96**, 59–91

Roe, H.M. 1968 *Medieval fonts of Meath*. Meath Archaeological and Historical Society.

Roe, H.M. 1975 The cult of St. Michael in Ireland. In C. Ó Danachair (ed.), *Folk and farm: essays in honour of A.T. Lucas*, 251–64. Dublin. Royal Society of Antiquaries.

Rouse, C. 1972 *Medieval wall painting*. Risborough. Shire Publications.

Sekules, V. 2001 *Medieval art*. Oxford University Press.

Stalley, R. 1987 *The Cistercian monasteries of Ireland*. London and New Haven. Yale University Press.

Steer, K.A. and Bannerman, J.W.M. 1977 *Late medieval monumental sculpture in the west Highlands*. Edinburgh. Royal Commission on Ancient and Historical Monuments of Scotland.

Wallace, J.N.A. 1936–41 Frescoes at Urlanmore Castle, Co. Clare. *North Munster Antiquaries Journal* **1–2**, 38–9.

Webb, J. 1824 Transactions of a French metrical history of the deposition of King Richard II. *Archaeologia* **20**, 1–423, pl. 4.

Went, A.E.J. 1952 Irish fishing spears. *Journal of the Royal Society of Antiquaries of Ireland* **82**, 109–34.

Westropp, T.J. 1911 Clare Island Survey: history and archaeology. *Proceedings of the Royal Irish Academy* **31** (1911–15), section 1, part 2, 1–78.

THE MUSICAL INSTRUMENTS DEPICTED AT CLARE ISLAND

Ann Buckley

ABSTRACT

This is an examination of the three vignettes of musicians on the ceiling vault of the Abbey at Clare Island: an organist and a lyre player on the north side and a harper on the south. A brief overview of the history of these instruments in an Irish context is followed by an attempt to interpret their significance in the iconographic programme of the paintings. Lyres and harps were well known in medieval Ireland and are frequently depicted on Irish religious monuments from the ninth and tenth centuries to the early fourteenth century. While the image of the organ is unique, documentary evidence suggests that the instrument was known in both rural and urban Ireland by the time the paintings were executed. In keeping with observations on other aspects of the paintings, it is concluded that the scenes of music-making are European in their symbolic content but Irish in most of their detail.

Introduction

There are three vignettes of musicians on the ceiling vault: an organist and a lyre player on the north side, and a harper on the south.

The organ (Pl. I)

The musician, coloured in red and supported on a yellow seat, apparently reaches towards the instrument with both arms outstretched. The organ is being played, but the image is too worn to identify whether it has one or more keyboards. The instrument is placed on a large red stand, perhaps a tripod (only three feet are visible). Its single rank of ten pipes, which alternate red and yellow, are in a somewhat uneven mitre formation, but the box or case in which they and the keyboard(s) would have been contained is not visible. It is thus a very stylised outline sketch, at least as it now survives.

In order to overcome the problem of two-dimensional representation the illustrator has placed the player and keyboard to the left of the pipes, i.e. at the side rather than the front of the instrument. To the right another figure, coloured in yellow, outlined in red and seated on a red stool or chair, operates a single set of (incised) bellows.

This is a unique representation in Irish iconography. To what extent, if any, might it reflect local experience?

The pre-Reformation history of organs in Ireland is poorly documented. Grattan Flood suggested that the instrument was already known in the country in the eighth century (Flood 1970, 31). There is no evidence for this, but the misunderstanding can be traced to an erroneous report on the presumed destruction of organs in a fire at the church in Cloncraff, Co. Roscommon, in the ninth century. The original source is apparently an entry for the year 814 in the Annals of Ulster, where a reference to *orgain Cluain Cremha* was glossed *direptio* in the margin. The gloss is a correct translation of the Irish term

Pl. I Organ, with person on left playing and person on right working the bellows.

orgain ('destruction'), but one which was misconstrued as referring to organs by a seventeenth-century scribe and quoted by numerous other writers ever since.[1]

The oldest documented organ in Ireland or Britain was a particularly large instrument built for Anglo-Saxon Winchester (McKinnon 1974). Organs existed in several English Benedictine abbeys by the end of the tenth century and were common in monastic churches in England and continental Europe from the eleventh century onwards. The instrument seems to have been well established in secular cathedrals from the end of the thirteenth century, even if its precise role in medieval liturgical services cannot be established (Williams and Owen 1988, 62ff). It was found in collegiate chapels and even in parish churches outside of London by the fifteenth century. The type represented here is a medieval positive organ, i.e., one placed on the floor (or on a stand, as here), which could be moved about, placed in the loft, on the screen, or in a side chapel. They were largely dependent on private endowments. Sometimes a church might have had more than one instrument, as their ambit was restricted, but larger organs were very rare before 1500.[2]

According to Grattan Flood there were organs at Christ Church Cathedral, Dublin, in 1358 and one in each of the Dublin cathedrals in 1450 (Flood 1910, 231), and a new organ was built in Christ Church in 1470.[3] He provided no documentary reference, and no sources are currently known that might shed light on these reports. Thus, while the organ may have been established in Ireland, or at least in Dublin, by the fourteenth century, the earliest verifiable evidence comes from the fifteenth century. The first apparently authentic reference to an organ in Ireland concerns an instrument in St Thomas's Abbey, Dublin, in the 1450s (see Holmes 1984, 1; Neary 1977, 20, n. 1).[4] Archbishop Tregury bequeathed his pair of organs[5] to the Lady Chapel at St Patrick's Cathedral in 1471 for use in the celebration of the Divine Office (Berry 1898, 26), and there are records of payments to organists at that establishment during the following two centuries (Grindle 1989, 133–4).[6]

In a note from the editor of Archbishop Tregury's will, it was stated that organ building 'flourished to a great extent' in Kilkenny in the fifteenth century (Berry 1898, 200). One John Lawless was said by Flood (1970, 100) to have been a celebrated Irish organ builder during the latter half of the fifteenth century, to have erected twenty organs in various parts of Ireland,[7] and to have been held in such high esteem as to be awarded privileges by Kilkenny Corporation, including a lease in 1476 on condition that he take up permanent residence in the city. The lease has been authenticated but not the other details.[8]

Holmes (1984, 1) also mentioned a Dominican, Fr John Roose,[9] who was active as a priest organ-builder in Kilkenny (and, according to Flood, also trained in organ-building there); he is said to have repaired two organs at York Minster, but Holmes gives no source for this. And another reference listed by Flood is attributed to the Annals of Duiske for the year 1460, according to which Brother Aengus, a Cistercian monk of Holy Cross Abbey, Co. Tipperary, came to Duiske to repair the 'old organ ... which, not having been used of late years, was sadly affected by damp, and the leather of the bellows was gnawed by rats.'[10] I have not succeeded in tracing this reference. However, some credibility is lent to it by the evidence we have for the existence of an organ at Duiske Abbey (referred to as 'the monastery in

Duske') at Graiguenamanagh, Co. Kilkenny, at the time of the dissolution of the monasteries. The account, from a document in the Public Record Office, London, is as follows:

> The monastery was dissolved, as Bective, q.v., 6 May 1536. The goods and chattels, including a bell still hanging in the belfry of the church, came into the hands of James earl of Ormond and Ossory, who is to account for them, except for the price (4.li. 13.s. 4.d.) of 5 cows and a pair of organs, accounted for by the accounting officer (White 1943, 197–8).[11]

Perhaps the most significant reference for the present topic, dealing as it does with the west of Ireland, concerns Thomas Bermingham, Baron of Athenry, and his wife, Annabella,[12] who, according to an account of the Dominican Abbey at Athenry, Co. Galway, are said to have granted 'three silver marks towards the building of the abbey-church organ in 1479' (Flood 1970, 100).

I have succeeded in tracing this reference via a seventeenth-century manuscript copy of older sources: the donation appears authentic, although I have been unable to verify the year[13] (a number of entries on the same folio are dated 1482, 1488 and 1491). As in the case of certain other references, this one seems to have been recorded twice—in the first record, the building of an organ is given as the purpose of the donation:

> Item praedicti dederunt ad fabricam organorum dicti conventus tres marcas de puro argento. (fol. 17r, no date)
>
> Item the aforementioned donated three marks of pure silver towards the building of the organs of the said convent.

in the second, mention is made of both the building and repair of the organ:

> Item praedicti dederunt ad fabricam cen [ceu] ad reparationem organorum dicti commentus [conventus] tres marcas de pura [puro] argento (f ol. 17v).[14]
>
> Item the aforementioned donated three marks of pure silver towards the building or repair of the organs of the said convent.

Another reference from western Ireland is provided by Bernard Adams, Bishop of Limerick from 1604 to 1626: in an appendix to the fourteenth-century Black Book of Limerick dated 20 August 1621, he reported on his restoration work on the cathedral following depredations wrought during the Elizabethan wars. In his report he mentioned that he acquired a very beautiful new organ, the older pair having been broken and destroyed ('antiquis duobus fractis et devastatis');[15] he also notes that he reorganised the choir (see MacCaffrey 1907, 165; also Flood 1970, 212; Boydell 1986, 550).[16]

All of the information gathered here suggests that organs were well known in Ireland by the period with which we are concerned and, furthermore, that they were not confined to larger, urban centres, but were used in rural abbeys as well.[17] It is also possible that privately endowed chapels in Ireland possessed such an instrument.

Whether the O'Malleys commissioned or paid for the upkeep of an organ in Clare Island Abbey we cannot establish. However, our most revealing references come from only slightly later in the fifteenth century than the suggested date of the painting. And since it is clear that landed families were involved in such patronage, this may be relevant for our understanding of the painting.[18] That being said, however, we ought not to overlook the possibility that the image is a 'cut' from another iconographic source, such as a Bible or a psalter or a stained glass window: in other words, that it is a borrowed image rather than one based on first-hand experience.

I am not aware of any other Irish iconographic source representing an organ. However, there are some English examples, including one on a single leaf from a lectern Bible from East Anglia, possibly Cambridge, dating to c. 1355.[19] It is little more than an outline sketch, though the keys are clearly visible (Bicknell 1996, 12, pl. 5). Another occurs in a stained glass image dating from the mid-fifteenth century from the Beauchamp Chapel, St Mary, Warwick, showing a larger, positive organ set in its case (Bicknell 1996, 24, pl. 6). In the latter instance, the angle of the performer in relation to the instrument is comparable to that of the Clare Island figure, as is the mitre formation of the main rank of pipes— although the mitre formation is more regular at Warwick, and the organ there includes two further sets of three long pipes at either side behind the main rank. A figure operating the bellows may just be discerned in the background. The fact that only one set of bellows is visible does not necessarily mean that two were not understood to be present. The same could apply to the Clare Island sketch.[20]

Pl. II Person playing a lyre.

The lyre (Pl. II)

The lyre, placed below the organ in an apparently separate scene, is a rectangular instrument with incised strings in fan formation. The strings are stretched over a soundbox, where they are attached to a clearly visible tailpiece fixed to the end of the instrument. Unfortunately the number of string incisions has not been recorded and such detail is not discernible from the photographic evidence, but six strings are likely. The upper part of the frame seems to slope towards the player, while the soundbox appears shorter at the player's end. While allowing for inaccuracy in precise detail, this nonetheless suggests that part of the base would have been hidden behind the player's right knee.

This musician, coloured in yellow, is seated on a red chair, facing east like the organist. The outline of his right arm can be seen in a realistic position for plucking the strings; his left hand, also in a realistic position, is behind the string band. The left hand may be plucking or perhaps

stopping the strings in order to produce more pitches.

A substantial repertory of lyre iconography survives on Irish high crosses and early medieval metalwork, all considerably earlier in date than this image: for example, the ninth/tenth century Barrow Valley crosses at Ullard, Castledermot and Graiguenamanagh; the Cross of Patrick and Columba at Kells; the Cross of Muiredach at Monasterboice; the Cross of the Scriptures at Clonmacnoise; the high cross at Durrow; the Kinitty Cross; and the Kells West Cross (see Buckley 1990, *passim*). But apart from the eleventh-century illustration on the Stowe Missal shrine (Buckley 1990, pl. XIV), no later depictions of plucked lyres are known in Ireland, which makes the Clare Island representation especially interesting. Typologically, this instrument, with its strings in splayed disposition, most closely resembles the Durrow Cross illustration. On the Durrow Cross the right hand is also in a playing position, whereas in most high cross representations it is placed on the frame of the instrument at the side nearest the player. A similar disposition of the knees to balance the lyre as that suggested in the Clare Island painting occurs at Durrow, Clonmacnoise, Kinitty and Ullard. The position of the left hand cannot easily be discerned in the stone carvings, but a parallel in an Irish manuscript painting may be found in the early tenth-century Cotton Psalter (British Library MS Cotton Vitellius F. XI, fol. 2r), albeit with a parallel string band and some rather fanciful detail (Henry 1960–61, pls IV and VI; Buckley 1995, 170, fig. 34). The best preserved example of this detail in an insular context is found in the Vespasian Psalter, an English manuscript thought to have been executed at Canterbury *c.* 725 (British Library, MS Cotton Vespasian A I, fol. 30v; see Buckley 1995, 163, fig. 28).

The lyre, known as a *cruit*, was the dominant court instrument in Ireland until superseded by the trilateral harp (to which the name was transferred) some time around 1000 AD. A *cruit* was used also to accompany the singing of the psalms and other sacred music, so there is no surprise in its presence on high crosses or, indeed, in a religious context at Clare Island. According to references in medieval Irish literature, monks were said to travel about with a small eight-stringed instrument (known as an *ocht-tédach*)

Pl. III Harpist.

attached to their girdle (Fleischmann 1952, 48). A three-stringed plucked lyre played by an ecclesiastic occurs on the shrine of the Stowe Missal. The restricted number of strings may be due to lack of space, but a three-stringed lyre-type instrument, called a *timpán*, which may perhaps be a small *cruit*, is also referred to in medieval Irish literature (see Buckley 1978, *passim*, and also 2001 for a brief summary).

The harp (Pl. III)

The instrument has a red frame relieved in yellow, with perhaps six red strings. The incised figure of the player, on a red chair with a yellow seat, wears clothing of now faded colour, with red sleeves and some yellow. An earlier account (Westropp 1911, 35) described the clothing as long yellow robes and identified six strings on the instrument, which is probably reliable, although this detail is no longer clearly visible. Actual harps of the time and earlier would have had a greater number of strings. The robust frame is characteristic of the late medieval Irish harp. The thick outward curve of the forepillar, the raised head and downward sweep of the arm are apparent. The base on which

the instrument rests can just be discerned, and an upward-sweeping diagonal red line suggests a characteristic, proportionately heavy soundboard. The musician's left arm is outstretched in a playing position.

Representations of medieval harps in Ireland are mostly of instruments of similar type but with a lighter frame, as, for example, that depicted in the hands of a cleric on the eleventh-century shrine known as Breac Maedhóg and the larger, light-framed instrument on the shrine of St Patrick's tooth (*c.* 1376; see Buckley 1990, pls XVI and XVII). Although the scarcity of surviving instruments makes it difficult to chart the history of the harp in detail, the heavier-framed harp is a later development: the oldest extant example—the so-called 'Brian Boru' harp, now housed in Trinity College Dublin—is believed to date to the fourteenth or fifteenth century.[21] The painting thus conceivably represents a contemporary instrument, with one notable difference. Normally Irish harps were supported on the left shoulder with the left hand playing the shorter, treble strings (i.e. those nearest the player) and the right hand the bass strings. That position is consistent with the instruments on the two shrines, but in the painting it is the left hand that reaches towards the lower end of the instrument. It may perhaps be due to lack of experience on the part of the illustrator, although we cannot, of course, be certain of any consistency or standardisation of practice. Furthermore, the vagueness of the position of the instrument in relation to the performer gives the impression that it is held quite separate from his body, almost aloft. It is not clear from the photographs whether this is due to weathering or to misrepresentation.

Context of the representations and symbolism of the musical scenes

Representations of lyre players on Irish high crosses are frequently accompanied by players of wind instruments—usually some kind of straight horn—as, for example, on the Cross of Muiredach, Monasterboice, and on the Durrow Cross. In addition, the Monasterboice cross includes a player of triple pipes. The musicians in these iconographic programmes attend upon Christ in Glory.

The Monasterboice scene is particularly suggestive as a thematic parallel because it also includes a representation of St Michael weighing the souls on a set of scales, which is placed just below the figure of Christ. The importance of St Michael in Irish spiritual devotion is discussed elsewhere (Morton, this volume, pp 103–105). His role in the Last Judgement scene at Monasterboice underlines the separation of the Just from the Wicked. The Just, placed to the right of Christ, are represented by a choir of monks singing to the accompaniment of musical instruments. A bird is perched on top of the lyre suggesting the inspiration of the Holy Spirit and, perhaps, emphasising the sweetness of the sound (a similar image occurs on the Stowe Missal shrine). The monks are singing pious and devotional music as a sign of their faith and goodness—an activity frequently referred to in medieval Irish literature to denote sanctity and also to express disapproval of the enjoyment of worldly pleasure. The scene at Clare Island Abbey might similarly be interpreted as a representation of heavenly music, associated with those who have led a good life, who thereby gain access to Paradise following the Last Judgement. It is an appropriate musical allegory for the soul of the deceased patron of the painting. It is interesting also to note that these musicians face east, towards the rising sun and the Second Coming of Christ.

The context for the harper is less immediately obvious. The reference may be a further nod to the patron, whose harper would always have been at his service to sing his praises and to entertain company during feasting.[22] With such an interpretation it might be fitting to place this figure among other royal pursuits such as hunting and sources of wealth such as fishing, which attest to the patron's status and the lavishness of his court. But, in addition, the harper may be enacting in an earthly context what the lyre-player and organist are doing in the life beyond—singing praise to God—thus representing the court and therefore the deceased's earthly existence in the best possible light.[23]

The representation of stringed instruments and their ability to calm nature has a long history in European art: the archetypal figure is Orpheus, whose attributes were adapted to a Christian context in late antique and medieval representations of David, Old Testament king and author of the Psalms. It is impossible not to regard the theme of David as an influence on Irish

representations, both literary and iconographic. However, as I have shown elsewhere (Buckley 2003, *passim*), Irish medieval artists seemed at times to have had a more earth-bound approach than their continental counterparts when introducing local interpretations of Christian themes. Rather than confining representations of music and musicians to the more formal David iconography found in English and French sources (with their images of a high-status, enthroned King David as an ideal type), they appear to have introduced local images of music-making, juxtaposing string players with the Crucifixion, Christ in Glory and the Last Judgement. Sometimes lyre players and harpers appear to narrate the story, perform laments and pious music or attend upon the surrounding company rather than stand apart in the more formalised, hieratic way of English and continental Davids: Irish representations show recognisable musicians engaged in activities that were familiar to the local society. This is further suggested by the inclusion of musicians in biblical scenes where no musician is mentioned in the Bible.

For example, in the scene showing the Miracle of the Loaves and Fishes on the head of the Cross of Patrick and Columba at Kells, a lyre player sits facing Christ as the food is distributed to the waiting multitude. Another example is the representation of the Judgement of Solomon on the carvings at Ardmore Cathedral, which include a harper in the same panel as King Solomon and the two women who each claim to be the mother of the infant. Although in the latter case Harbison has concluded that two separate scenes have been merged into one—the only plausible interpretation in strictly Biblical terms (Harbison 1995, 97)[24]—it could be viewed as an Irish interpretation of Solomon's court, where the king's harper would have been in attendance. By the same token, the Kells image indicates an assumption of the presence of musicians on the occasion of a feast.[25]

One is reminded also of the extent to which medieval musical imagery makes reference to rejoicing and to singing God's praises. Interestingly, one psalter with Irish connections provides a possible link, whether direct or otherwise, with the apparently late depiction of the Clare Abbey lyre. The Christ Church Psalter contains a historiated initial at the opening of Psalm 80, 'Exsultate Deo', showing a group of monks singing and playing instruments, including a hybrid stringed instrument that is held and played like a harp but looks like a poor representation of a quadrilateral lyre with a very thick frame, suggesting that the painter did not quite know what he was doing. The manuscript was prepared in England in the late fourteenth century for Christ Church under the orders of Prior Stephen de Derby: given the otherwise inexplicable nature of this instrument, it may be that the artist was attempting to represent an Irish instrument and was unfamiliar with its details.[26] This and the Clare Abbey lyre may conceivably point to a more widespread use of such imagery in Irish manuscript and wall paintings that have since been lost or may still await recovery.

These observations on the musical imagery of the paintings support the thesis put forward elsewhere in this volume, namely that the iconographic programme contains themes that were at the same time European in concept and Irish in much of their detail.

Acknowledgements
I am grateful to Christoph Oldenbourg for kindly supplying me with photographs of the images under discussion here. I am also indebted to him for responding to my questions concerning various aspects of the iconographic detail of the paintings. I thank Conleth Manning (National Monuments Service, Department of the Environment, Heritage and Local Government) for several valuable suggestions made in the course of completing this paper; Andrea Clarke (Department of Manuscripts, The British Library) for assistance with Latin texts; James McEvoy (NUI Maynooth), Anthony Harvey, Jane Power and Angela Malthouse (Royal Irish Academy) for further advice and for assistance with refining the translation. For their generous assistance with a number of bibliographic and related matters I thank Barra Boydell (National University of Ireland, Maynooth), John Bradley (National University of Ireland, Maynooth), Christopher de Hamel (Parker Library, Corpus Christi College, Cambridge), Mark Fitzgerald, Conleth Manning, Michael O'Neill; and Stuart Ó Seanóir (Department of Manuscripts, Trinity College Dublin).

NOTES

1. For a complete account see Fleischmann 1952, 48.
2. See Bicknell 1996, 11–25, for a well-documented account of the period *c.* 900–1500 in England.
3. According to Grindle (1989, 243, n. 26), the first and third of these references are included in an article of Flood's entitled 'The organs of Christ Church Cathedral, Dublin', which was pasted into a volume of the Cathedral's chapter acts (vol. 22, October 1616). Grindle reports that it had been cut out of the publication in which it first appeared, without any reference as to the source.
4. Dr Barra Boydell kindly consulted his copy of Holmes's unpublished pamphlet on my behalf. He reports that Holmes cites J.T. Gilbert as his source while noting that the document containing the reference was destroyed in the Public Record Office fire of 1921. Unfortunately Holmes does not indicate which of Gilbert's many publications he consulted, and it remains untraced so far.
5. The instrument was often referred to historically in this way. Although the precise reason is not clear,

it is thought that it might refer to two ranks of pipes, providing two registers, or perhaps to the presence of more than one keyboard—resulting in two instruments, as it were.
6. For further discussion, see Boydell 1986, 548; Neary 1997.
7. Two in Christ Church in 1470, according to the account reported by Grindle. See note 2 above.
8. See Watters 1872–3, 543, which contains a reference to 'John Lawles orgon maker'.
9. Flood 1970, 99, refers to him as 'Brother John Rouse'.
10. John Bradley has alerted me to the existence of a bogus work of this name by John and Patrick O'Leary from *c.* 1890, which, he says, 'has been misleading people ever since'. As far as the real Annals of Duiske are concerned, only two manuscript copies survive, both from the seventeenth century. They are included in two independent series of extracts from the Duiske Chartulary, now preserved in TCD MS 578, fols 13–18, and BL MS Lansdowne 418, fols 61–4,

NOTES

respectively. The Dublin manuscript was prepared apparently for Archbishop Ussher and is in his hand; the London copy was compiled under the direction of Sir James Ware (Nicholls 1983, 92ff). Neither has a reference to Brother Aengus nor to an organ at Duiske. I am indebted to Stuart Ó Seanóir for kindly consulting TCD MS 578 on my behalf.

11. I thank Conleth Manning for drawing this reference to my attention. White states in his introduction (p. vii) that the photostats from which he compiled his account were deposited in the National Library of Ireland.

12. Spelled 'Anabla' in the source; Bermingham is rendered 'Bremingham' (see note 13 below).

13. Flood's reference to BL MS Sloane 4784, 43, n. 4, is incorrect. It is in fact MS Additional 4784, f. 17r–v (olim 52r–v). Thomas and Annabella are recorded as having given numerous donations of land as well as cash for the maintenance of the abbey, for memorial services on the anniversary of the death of John, their son and heir (a marginal note gives 1488 as the year of his death), and for the celebration of the Divine Office (quadraginta marcas usualis monetae quarum obtentu dictus conventus emit ad cultum divini officii capam et casulam cuam [cum] duabus tumiculis rubei coloris de puro serico) 'forty marks of common coinage, on receipt of which the said convent bought a cope and a chasuble, with two red cords [girdles?] of pure silk, for the celebration of the Divine Office'. In the opinion of O'Grady (1926, vol. 1, 20), who comments on the repetitiveness of the entries, this manuscript represents a quantity of 'rough material' amassed for the use of the 'Four Masters' when compiling the annals. It contains excerpts from various collections of annals covering the period from the fifth to the sixteenth centuries. I have found no reference to the donation in the published versions of the Annals of the Four Masters. However, the death of Thomas MacF[h]eorais (Bermingham), Lord of Athenry and of Conmaicne of Dunmore is recorded for the year 1473. It is stated that his son, Thomas Oge, took his place but that the title was given to the son of Richard Mac F[h]eorais 'in opposition to him' (O'Donovan (ed.) 1848, vol. 1).

14. I am indebted to Dr Andrea Clarke for assistance with deciphering the Latin text and to Professor James McEvoy, Dr Anthony Harvey, Jane Power and Angela Malthouse for further advice on the text and for assistance with refining the translation.

15. 'duobus' probably refers to a single 'pair' of organs although, even if unlikely, one cannot say for certain that it might not suggest the prior existence of two distinct instruments.

16. The Black Book is now housed in the Limerick [R.C.] Diocesan Archives. Flood (1910, 231) also referred to the existence of an organ in Limerick in 1450. We cannot substantiate this, but Adams's account lends further credence to the possibility.

17. When this article had reached an advanced stage in the editorial process, Conleth Manning informed me of another pre-Reformation reference to an organ in the possession of an Irish Abbey. The Franciscan friary of Killeigh, Co. Offaly, which was in a Gaelic lordship, was sacked by Lord Leonard Grey, the Lord Deputy, in a foray during November 1537. The latter is quoted as follows: 'and oute of the seid abbye I broght a peyer of orgons, and other nesessarie thinges for the kynges collage of Maynoth, and as muche glas as glassid part of the windous of the chyrche of the seid Collage, and much dall of the windous of his Graces castell of Meynoth'.

The reference is in an anonymous article on Maynooth Castle published in the *Journal of the Royal Society of Antiquaries of Ireland* 44 (1914), 281–94, apud 285. The original source is not provided.

18. Given the necessarily unsystematic and random nature of the work involved, there may well be more such information awaiting discovery. I should be immensely grateful for any further leads on the use of organs, on organ-building, and on patronage in pre-Reformation Ireland, or indeed on any aspect of music history that may emerge in future archival and art-historical research.

19. The leaf was sold at Sotheby's on 19 June 1990 and remains in private ownership. I am indebted to Dr Christopher de Hamel for this information and related details.

20. See also Bowles 1982 for numerous examples of organ iconography from the fifteenth century.

21. See Rimmer 1984 for a representative survey.

22. A unique memorial to an Irish harper exists in the form of the early sixteenth-century tomb effigy of a harper and his wife at Jerpoint Abbey, a Cistercian site. It is thought to represent a well-respected minstrel of the Earls of Ormond. The remains of the inscription suggest the name William O'Houlahan (see Hunt 1974, vol. 1, 175–6; vol. 2, pl. 181).

23. There are many exemplary tales in medieval Irish literature that rely on the idea of 'good' and 'bad' musicians to make a moral point. The 'good' musicians play pious music, the 'bad' ones are associated with excessive drinking and other socially unacceptable behaviour. For further discussion, see Buckley 1997, *passim*.

NOTES

24. Harbison's article represents a thorough exploration of the Ardmore scenes in terms of psalter illustration, though he is far from convinced that they were derived from Irish sources (Harbison 1995, 100). This does not undermine my interpretation, but may suggest some element of local adaptation of imported imagery.

25. For further discussion, see Buckley 2003, *passim*.
26. The psalter is now Bodleian Library, Oxford, MS Rawl. G. 185; Psalm 80 occurs on f. 68v. See Boydell 1999, 25–6 and the frontispiece illustration of the same book.

REFERENCES

Berry, H.F. 1898 *Register of wills and inventories of the diocese of Dublin in the time of archbishops Tregury and Walton 1457–1483*. Dublin. Dublin University Press for the Royal Society of Antiquaries of Ireland.

Bicknell, S. 1996 *The history of the English organ*. Cambridge. Cambridge University Press.

Bowles, E.A. 1982 A preliminary checklist of fifteenth-century organs in paintings and manuscript illuminations. *The Organ Yearbook* **13**, 5–30.

Boydell, B. 1986 Music before 1700. In T.W. Moody and W.E. Vaughan (eds), *A new history of Ireland*, IV, 542–628. Oxford. Clarendon Press.

Boydell, B. (ed.) 1999 *Music at Christ Church before 1800: documents and selected anthems*. Dublin. Four Courts Press.

Buckley, A. 1978 What was the tiompán? A problem in ethnohistorical organology: evidence in Irish literature. *Jahrbuch für musikalische Volks- und Völkerkunde* **9**, 53–88.

Buckley, A. 1990 Musical instruments in Ireland from the 9th to the 14th centuries: a review of the organological evidence. In G. Gillen and H. White (eds), *Irish musical studies 1*, 13–57. Blackrock. Irish Academic Press.

Buckley, A. 1995 Music-related imagery on early Christian insular sculpture: identification, context, function. *Imago Musicae/International Yearbook of Musical Iconography* **8** (1991), 135–99.

Buckley, A. 1997 Music and manners: readings of medieval Irish literature. *Bullán: an Irish Studies Journal* **3**(1), 33–43.

Buckley, A. 2001 Timpan/Tiompán. In S. Sadie (ed.), *The New Grove dictionary of music and musicians*, vol. 25, 482–3. London. Macmillan.

Buckley, A. 2003 Representations of musicians in medieval Christian iconography of Ireland and Scotland as local cultural expression. In Katherine A. McIver (ed.), *Art and music in the early modern period. Essays in honor of Franca Trinchieri Camiz*, 217–31. Aldershot. Ashgate.

Fleischmann, A. 1952 References to chant in early Irish MSS. In S. Pender (ed.), *Féilscríbhinn Torna*, 43–9. Cork. Cork University Press.

Flood, W.H. Grattan 1910 Irish organ-builders from the eighth to the close of the eighteenth century. *Journal of the Royal Society of Antiquaries of Ireland* **40**, 229–34.

Flood, W.H. Grattan 1970 *A history of Irish music*. 3rd edn. Shannon. Irish University Press.

Grindle, W.H. 1989 *Irish cathedral music: a history of music at the cathedrals of the Church of Ireland*. Belfast. Institute of Irish Studies, Queen's University Belfast.

Harbison, P. 1995 Architectural sculpture from the twelfth century at Ardmore. *Irish Arts Review* **11**, 96–102.

Henry, F. 1960–61 Remarks on the decoration of three Irish psalters. *Proceedings of the Royal Irish Academy* **61**C, 23–40.

Holmes, J. 1984 The organ in Ireland. Unpublished pamphlet.

Hunt, J. 1974 *Irish medieval figure sculpture 1200–1600*, 2 vols. Dublin and London. Irish University Press / Sotheby Parke Bernet.

MacCaffrey, J. (ed.) 1907 *The Black Book of Limerick*. Dublin. M.H. Gill & Son.

McKinnon, J. 1974 The tenth-century organ at Winchester. *The Organ Yearbook* **5**, 4–19.

Neary, D.M. 1997 Organ-building in seventeenth- and eighteenth-century Dublin, and its English connections. *The British Institute of Organ Studies Journal* **32**, 20–7.

Nicholls, K.W. 1983 Late medieval Irish annals: two fragments. *Peritia* **2**, 87–102.

O'Donovan, J. (ed.) 1856 *The Annals of the Kingdom of Ireland by the Four Masters* [Annála Ríoghachta Éireann], 7 vols. Dublin. Hodder, Smith and Co.

O'Grady, S.H. 1926 *Catalogue of Irish manuscripts in the British Museum*, vol. 1. London. Printed for the Trustees of the British Museum.

Rimmer, J. 1984 *The Irish harp*. Cork. Mercier Press.

Watters, P. 1872–3 Original documents connected with

THE ABBEY AND ITS CONTEXT

THE ABBEY IN ITS LATER GOTHIC CONTEXT

Roger Stalley

ABSTRACT

Thanks to the recent programme of conservation, it is possible to look at the architecture and the decoration of the church on Clare Island in a wider context. The circumstances in which the church was founded remain puzzling, but it is likely that it was linked to an agreement between the O'Conor kings of Connacht and the O'Malleys. By the later middle ages the church had evidently become more a monument to the chiefs of Clare Island rather than a spiritual outpost of the Cistercian order. The imagery employed by the painters—hunting scenes, warriors, musicians—has a secular flavour, closely associated with the theme of lordship. The choice of images is examined in the context of late Gothic art, both in Ireland and further afield, and analogies are made with the tombs of Scottish lords in the Western Isles. The painting of simulated ribs on the underside of a barrel vault may represent a practice that was once more common in Ireland. The key to the date of the painting lies with the design of the canopied tomb, a type that can be found in the west of Ireland from c. 1385 to c. 1561. The tracery can be related to patterns employed in the north of England and Scotland, but the overall design finds its closest parallel at Strade (Co. Mayo), a monument that recent scholarship has assigned to the first half of the sixteenth century.

Introduction

The so-called 'Abbey' on Clare Island is an enigmatic and puzzling monument. Although the church belonged to the Cistercian order, there is nothing obviously monastic about its architecture or its decoration, and the paintings in the chancel, which have attracted the interest of antiquarians for more than a century, seem curiously at variance with monastic ideals. By the late 1980s these paintings were deteriorating so rapidly that many visitors to the island feared they might soon be lost for ever. Even when the programme of consolidation and conservation got under way in 1992, nobody predicted that new and previously unknown images would be revealed. The uncovering of the mounted warrior on the south wall was thus one of the most unexpected and exciting artistic discoveries in recent times. While our knowledge of the paintings is likely to remain frustratingly incomplete—for no amount of investigation will compensate for the permanent loss of painted plaster—we now have a far better idea of the overall scheme than was thought possible twenty years ago.

For many years the struggle to identify the images, only dimly visible on the vault of the chancel, overshadowed all other items of discussion, but thanks to the programme of conservation and the careful cataloguing of the paintings it is possible to look at both the architecture and the decoration in a wider context.

Pl. I   Canopied tomb in the chancel of the Abbey, Clare Island (DOEHLG).

There are a number of questions that arise, albeit ones that cannot be answered with any certainty. One immediate issue is the degree to which the scheme encountered on Clare Island was unique in late medieval Ireland. The majority of medieval churches—chapels, parish churches, abbeys and cathedrals—have come down to us as ruins and it is likely that most, if not all, were once bedecked with painted images. The work at Clare Island may thus provide useful hints about what has been lost elsewhere. There are good reasons for assuming that the painting was carried out by local craftsmen, but were they pursuing local methods or following standard artistic conventions of late Gothic Europe? This raises the question of how far Irish artists, particularly those active in the Gaelic lands west of the Shannon, developed their own versions of the Gothic style. Although some of the images on the vault have Christian connotations, the extent to which the choice of subjects was dictated by coherent religious themes remains far from clear. Many observers have been puzzled by the curious mixture of subjects, a confusing *mélange* that seems out of place in a building belonging to the

Cistercians, a point that raises questions about the function of the church and the nature of its relationship with the monks at Abbeyknockmoy. Then there are the questions posed by the spectacular canopied tomb in the chancel (Pl. I), with its flowing Gothic tracery; this is one of several such tombs in Connacht, the design of which may offer important clues to the function of the chapel and the date of the paintings.

The church and its function

Thanks to the investigations of Conleth Manning, it is clear that the original church was built in the early thirteenth century under the auspices of Abbeyknockmoy (Co. Galway) (Pl. II). The monument has traditionally been described as an abbey, but it seems unlikely that it ever functioned as a normal Cistercian monastery. Admittedly, seventeenth-century records speak of the 'late dissolved abbey of Cleere' as if there was an active community of monks on the island, but one searches in vain for any sign of the traditional cloister and domestic buildings.[1] Nor was the amount of land that went with the church (half a quarter) sufficient to sustain more than a small community.[2] So what exactly was the status of the church? James Ware described it as a 'cell' of Abbeyknockmoy,[3] a somewhat ambiguous term that historians have tended to use rather loosely to describe a variety of sites, including failed houses, granges and local churches in monastic ownership.[4] The founding of small cells for two or three monks was contrary to Cistercian practice, though one scholar has suggested that Clare Island provided a place of retreat for the monks, who used it for 'a change of air, for health purposes, and perhaps also for greater retirement'.[5] Others have suggested that the site functioned as a grange, in effect being maintained as an isolated farmstead for the benefit of the community. In cases where granges lay at some distance from the mother house, it was necessary to build domestic quarters for the lay brothers and workers, along with barns, stockyards and farm buildings. But there is no evidence for this on Clare Island. Had the site functioned as a grange one might expect this to be reflected in local place names.[6] Although the Cistercians are renowned for their skill as farmers, the order soon began to lease its possessions, especially when the lands concerned were far removed from the monastery

Pl. II   Abbeyknockmoy, general view of the monastic ruins.

itself. It is perhaps more likely that, from the start, the land was leased to one of the islanders.

In the absence of any other medieval church on the island, the surviving church must have served as a place of worship for the local community.[7] Although the statutes of the Cistercian order forbade the ownership of parish churches and dependent chapels, this rule was widely ignored by the thirteenth century. Was the parcel of land on Clare Island given to Abbeyknockmoy on the understanding that the profits or rents were to be used to maintain a chaplain? Whatever the circumstances, it seems likely that the building functioned along the lines of a parish church, the advowson of which belonged to Abbeyknockmoy. At first sight it seems curious that the donation was presented to a monastery almost 50 miles away. But it is worth noting that the O'Conor kings of Connacht were among the chief benefactors of Abbeyknockmoy,[8] the site of the abbey being just a few miles south of Tuam, one of the kings' principal residences (Pl. II). In fact, Cathal Crobderg Ó Conchobair spent his last days in the Cistercian abbey, dying there in 1224 in the habit of a monk.[9] The association between the O'Conors and Abbeyknockmoy on the one hand,

and Abbeyknockmoy and Clare Island on the other, suggests there was a 'political' dimension to the foundation of the church on Clare Island, the result perhaps of some 'arrangement' between the O'Malleys and the O'Conors, the nature of which we can no longer discern.[10] Although the church must have served the local populace, there is no evidence that Clare Island was ever formed into a separate parish. From at least the sixteenth century the building formed one of several outlying chapels within the parish of Kilgeever (Louisburgh).[11] It is remarkable that the link with the Cistercians endured throughout the middle ages, with the site still being listed as part of the property of Abbeyknockmoy in 1584. Furthermore, the painted ribs in the chancel provide a striking reminder of the enduring nature of this relationship, for they are clearly modelled on the real ribs in the presbytery of Abbeyknockmoy (Pl. III).

The vaulted chancel

For approximately two hundred years the church existed as no more than a straightforward single-cell building, covered by timber rafters and (probably) stone slates. It was only in the later

Pl. III   Abbeyknockmoy, the ribbed vault in the presbytery.

middle ages that the chancel was added, along with the impressive barrel vault. The introduction of stone vaulting in the medieval period is generally looked at in practical terms, as an insurance against fire, but this is only part of the story. Vaults were far more than a structural asset: in many cases they identified the most sacred part of the building, with the stone vault acting like a canopy or baldachino over the sanctuary and altar. The formula, found in churches large and small throughout the middle ages, is one encountered in many an Irish Cistercian monastery, as at Abbeyknockmoy or Boyle. Stone vaults were thus a way of defining the most sacred spaces within the church; in this respect the builders at Clare Island were following long-established medieval practice.

More interesting is the fact that the chancel was covered with a barrel vault (see Morton, this volume, pl. V, p. 102) rather than some species of ribbed vault. Semicircular-shaped barrel vaults are generally associated with Romanesque building in the eleventh and early twelfth century. Although still used in utilitarian or secular

structures, they became quite rare in Gothic church design. In fifteenth-century Ireland, however, the barrel vault enjoyed a modest revival, with new examples being erected over the chancels of the Cistercian churches at Jerpoint and Kilcooly. In a European context, it seems a curiously retrograde step, not least since the barrels had a semicircular rather than a pointed section, contrary to what is expected in Gothic architecture. Barrel vaults were also erected in more modest churches, as at Liathmore (Co. Tipperary), and they are common in mendicant architecture (a striking example survives in the chapter house of the Franciscan friary at Ennis). Compared with a ribbed vault, a barrel vault was cheaper and simpler to construct, and this no doubt was part of the attraction. But to exert any architectural impact, the surface needed some form of decoration.

Mindful no doubt that a ribbed vault was a more prestigious form, the painters at Clare Island indulged in a remarkable piece of architectural simulation, with painted ribs creating the illusion of a far more sophisticated piece of architecture.

It is worth noting that the form of the ribs is not like those of the standard Irish ribbed vault of the fifteenth century. They are dark in colour (as if suggesting carboniferous limestone), and they are devoid of mouldings or chamfered edges. Broad ribs of this type are known from only two monuments in Connacht, the churches at Abbeyknockmoy and Ballintober, both dating from the first quarter of the thirteenth century (Pl. III). It is hard to avoid the conclusion that the painters were asked to copy those in the chancel of Abbeyknockmoy, suggesting that the link with the Cistercians was more than a formality. As the vault at Abbeyknockmoy was erected about 1220, the painters were copying a structure that was two centuries old, an extraordinary piece of antiquarianism. It seems that the painted ribs were not just a convenient compositional device: both the patron and the painters must have been well aware that ribs would endow their tiny chancel with greater status, while at the same time providing an explicit link with the 'mother' church at Abbeyknockmoy.[12]

If a remote chapel like that on Clare Island was provided with simulated ribs, it seems likely that barrel vaults in far wealthier churches, like those at Jerpoint, Kilcooly or Boyle for example, were given a similar treatment. It also suggests that when genuine rib vaults were erected, as at Abbeyknockmoy and Ballintober, the intervening cells were ornamented. Given the fact that the painter (or painters) who carried out the work on Clare Island must have been familiar with the chancel at Abbeyknockmoy, it is quite possible that they were recruited via the mother house. Outside Ireland the painted rib was quite common, for it was a useful way of livening up the plain surfaces of a groin or ribbed vault. At Loudun in the Loire valley, for example, painted ribs were part of a colourful fourteenth-century design used to embellish a Romanesque barrel vault over the chancel. Here the ribs are treated in a highly decorative fashion, a contrast to the very explicit architectural form of those on Clare Island.

The wall tomb

The most prominent feature of the chancel at Clare Island is the extravagant tomb in the north wall, furnished with a tracery screen and ogee-headed gable (Pl. I). The location of the tomb is one that was usually reserved for founders or major benefactors, so the monument must commemorate the chief of the island, presumably one of the O'Malleys.[13] As explained in earlier chapters, the tomb must have preceded the painting on the adjacent walls, since the plaster abutted the stonework and the paintings continued onto it; in fact the monument may have belonged to the individual who paid for the addition or remodelling of the chancel, the act that justified the creation of a 'founder's tomb'.

The wall tomb has a long history, stretching back to the *arcosolia* or wall niches found in the catacombs of Rome. By the second half of the thirteenth century elaborate tombs with a chest below and canopy above became fashionable in both France and England. Simple wall tombs, defined by cusped arches, soon found their way to Ireland, and there is a fine series in the parish church at Gowran (Co. Kilkenny). But two centuries elapsed before ornate canopied tombs appeared in an Irish context, the majority of them being located in Connacht. The example at Clare Island lies firmly within this western group. Distinguishing features include an arcaded tomb chest surmounted by a screen of open tracery under a single enclosing arch. At the base of the design there is usually a semicircular or segmental arch, furnished with hanging cusps. Despite the wide range of designs found in England, there appear to be no exact precedents for this very distinctive Irish approach.[14]

On Clare Island the chest itself is decorated with somewhat squat panels of tracery in the English Perpendicular style; there are six mini-arches, each with ogee heads. The canopy is provided with flowing tracery, completely detached from the wall behind: at the base of the design, the semicircular arch with hanging cusps supports a central, vertical mullion. Along the sides are huge, elongated leaf motifs, also furnished with cusps. The latter are so large that they almost meet in the middle, forming circular motifs, an unusual and rather ungainly arrangement. It is important to remember the two oddities in the construction of the monument, already noted by Conleth Manning, namely the discrepancy between the dimensions of the chest and those of the canopy and the difference in stone types. The chest is made of local serpentinite, the canopy of carboniferous limestone. The limestone was evidently brought from the mainland, presumably from an established quarry or workshop where the pieces were carved. The dimensional inconsistencies may well be explained by the importation of the canopy in a prefabricated state. There is, in fact, no proof that the two parts of the tomb were designed as a single scheme; the canopy might have been an afterthought.

The date of the canopy is crucial: if this can be established, we have a useful clue to the date of the paintings, and perhaps even to the building of the chancel itself. There are at least seven other canopied tombs in Ireland, located at Dungiven; Sligo; Kilconnell (two); Strade (Pl. IV); Athenry; and Galway (at the church of St Nicholas).[15] The tomb at Dungiven has traditionally been associated with Cooey na Gall O'Cahan who died in 1385, and though the tomb has usually been dated to a later period, the stylistic details are not inconsistent with a late fourteenth-century date, as McNeill has argued.[16] There is more certainty about the date of the O'Criain monument at Sligo, which carries an inscription referring to the year 1506. The design of neither of these tombs, however, has much bearing on Clare Island. More closely related are the two fine monuments in the

friary at Kilconnell. The tomb in the chancel there has magnificent geometrical tracery (Pl. V); in the context of Clare Island, it is worth noting the wide semicircular arch with hanging cusps, as well as the Perpendicular arcading on the chest. Harold Leask wondered whether the Kilconnell tomb might belong to Malachy O'Kelly who died in 1464, but he dismissed this idea in the belief that the style indicated a later date.[17] The second tomb at Kilconnell, this time located in the nave, is furnished with curvilinear tracery: the ogee frame recalls that at Clare Island, as does the lower arch, but the tracery above is entirely devoid of cusps, creating a more dynamic flowing movement. The figure sculpture on the chest appears to belong to the late fifteenth or early sixteenth century (*c.* 1470–1530).[18]

In terms of overall composition the most impressive of the Connacht tombs is that at Strade (Co. Mayo) (Pl. IV).[19] Whoever was commemorated in this case (again the identity is unknown) was an avid pilgrim, to judge from the sculptures on the tomb chest.[20] The carvings include details

Pl. IV   Strade, canopied tomb in the choir of the Augustinian friary (DOEHLG).

Pl. V   Kilconnell Friary, canopied tomb in the chancel.

that link the work to the so-called 'Ormond school', active in the Kilkenny region *c.* 1500–40; in fact the image of Christ showing his wounds replicates the Kilkenny designs almost exactly. The outstanding work of the Ormond 'school' is the tomb in Kilkenny Cathedral of Piers Butler, the eighth earl of Ormond, who died in 1539,[21] and it is difficult to believe that the sculpture at Strade is all that different in date, a point that has obvious implications for the work on Clare Island. It is tempting to regard the tomb on the island as a simplified (and slightly debased) version of that at Strade. Closely related to Strade is the so-called 'Joyce' monument situated in the south transept of the church of St Nicholas at Galway, a part of the fabric that, it is believed, was not erected until 1561.[22] This, the most ornate and densely patterned tomb in the group, suggests that the canopied tomb remained in fashion for a period of almost two hundred years (*c.* 1385 until *c.* 1561).

The tracery employed at Galway, Strade and Clare Island has sometimes been compared with French 'flamboyant' designs, the idea being that influence from French architecture percolated to the west, perhaps through the port of Galway. But this is to ignore the fact that flamboyant tracery was also fashionable in Scotland and the north of England in the years around 1500. Identical motifs can be seen on a wooden screen at Carlisle cathedral (1484–1507) (Pl. VI), and a wide range of curvilinear patterns is included on woodwork at King's College Chapel Aberdeen (1506–9).[23] The designs at Carlisle have an uncanny resemblance to those on the Irish tombs, a reminder of the important role that woodwork, long since vanished, must have played in the furnishing and design of Irish medieval churches.

Pl. VI   Carlisle cathedral, detail of the Gondibur screen 1484–1507.

The equestrian warrior

In contrast to the situation elsewhere, the canopied tomb at Clare Island dominates the small chancel, conveying the impression that the building is more a chantry chapel than a monastic sanctuary, a place for the celebration of lordly power rather than Cistercian spirituality. This impression is heightened by the painting of the equestrian warrior on the wall opposite the tomb (Morton, this volume, pl. II, p. 99). The image of the mounted knight is ubiquitous in medieval art —it is found, for example, on countless medieval seals, the very embodiment of aristocratic authority.[24] On occasions it is even found in direct association with funerary monuments, as with the carvings on the tombs of Edmund Crouchback (d. 1296) and Aymer de Valence (d. 1324) in Westminster Abbey.[25] Where the design at Clare Island differs from conventional images is in the dress of the warrior and the fact that he wields a spear rather than a sword. The rider is apparently wearing a pointed helmet or *basinet*, but the rest of

Pl. VII   de Burgo knight *c.* 1584 (The Board of Trinity College Dublin).

Pl. VIII   Art Mac Murrough meeting the Earl of Gloucester (British Library).

his dress is not so clear. He appears to be protected by a long hauberk of mail, the uniform of the gallowglasses as depicted in sculpture on tombs at Roscommon and Dungiven and repeated on the splendid effigy of a de Burgo (Burke) knight at Glinsk (Co. Roscommon).[26] However there is no indication of the undergarment or *aketon* that usually protrudes below, a garment that was essential for anyone wearing a coat of mail. It is well known that Gaelic horsemen showed no interest in saddles or stirrups, and in this respect the Clare Island figure follows Irish custom.[27] One would imagine that the wearing of a coat of mail without a saddle would be uncomfortable (at least for the horse), but the sixteenth-century image of a de Burgo knight confirms that this was normal procedure, the hauberk being split at front and back almost to the waist (Pl. VII).[28] It is worth recalling what the Spanish traveller, Count John de Perilhos, had to say about Irish knights in 1397, when he made his visit to the court of O'Neill.[29]

He has forty horsemen, riding without saddle on a cushion, and each wears a slashed cloak; moreover, they are armed with coats of mail, and wear them girded, and they have throatpieces of mail and round helmets of iron, with swords and sword-blades and lances very long, but very thin in the manner of the ancient lances, and they are two fathoms long.

Whatever the exact dress depicted at Clare Island, the painting represents a rare and valuable portrait of a Gaelic lord in full cry. There are not many comparable images. The fourteenth-century shrine known as the Domhnach Airgid includes depictions of charging horsemen, but in this case the riders are wielding swords and wear what have been interpreted as quilted aketons (along with extraordinary hats or *chapels-de-fer*).[30] Much closer to the Clare Island representation is the famous miniature of Art Mac Murrough racing down the hills to meet the earl of Gloucester, one of the French illuminations that accompany Jean Creton's account of Richard II's expedition to

Pl. IX   Annagh (Kerry) relief carving of a horseman.

Ireland in 1399 (Pl. VIII).[31] Creton admired Mac Murrough's horsemanship, not least since the horse was 'without housing or saddle'. In his hand Mac Murrough 'bore a great long dart, which he cast with much skill'.[32] The most puzzling aspect of the image of Art Mac Murrough is that it was painted by a professional illuminator in Paris, who presumably never visited Ireland; nonetheless he has managed to convey some of the more characteristic Irish details.[33] In Irish society the spear evidently ranked as highly as the sword, a point reflected in a passage from a late thirteenth-century ode to Hugh O'Conor, king of Connacht (1293–1309) that describes the apparel of the king; 'the gold along the edges of his lance covers much of this broad-headed thrusting spear'.[34] There are several images of spear-wielding horsemen both in Ireland and in the Western Isles of Scotland. A little-known medieval relief built into the walls of a ruined church at Annagh (Co. Kerry) shows a rider with peculiarly rigid legs holding what appears to be a spear (Pl. IX).[35] Perhaps more relevant is a fifteenth-century tomb slab at Iona depicting a mounted warrior, also armed with a spear.[36] As far as one can judge, Irish native armour changed little between the fourteenth and the sixteenth century, and Gaelic lords apparently showed no interest in the plate armour depicted on the effigies of their Anglo-Irish counterparts.

Pl. X   Grave slab of Murchadus Macduffie (d. 1539), Oronsay Priory (Argyll) (Royal Commission on the Ancient and Historical Monuments of Scotland).

Hunting and other imagery in the paintings

By far the most popular pastime for the aristocracy throughout Europe was hunting, a pastime the lords of Clare Island appear to have shared.[37] Even so, it is surprising to find a hunting scene depicted in such close proximity to the monument, with a stag fleeing between the pinnacles of the tomb itself. While the juxtaposition is bizarre, the representation of hunting in medieval churches is not. A hunt is portrayed in the north transept of the Cistercian church at Holycross, and another example can be found in the nave of the parish church at Hailes (Gloucestershire), the *capella extra portas* of the great Cistercian abbey nearby. In the latter case the collection of subjects on the walls is almost as confusing as that at Clare Island: heraldic shields, grotesque monsters, popular saints like St Christopher and St Michael (complete with the scales) appear along with an extensive hunting scene. Hunting was also a favourite choice for misericords, and it is even encountered on floor tiles.[38] More significant for Clare Island, perhaps, is the association between hunting and funerary sculpture, an association that is made explicit in the monuments of western Scotland. The grave slab of Murchadus Macduffie (d. 1539) at Oronsay Priory (Argyll), for example, includes a hunting scene with three deer and four dogs, one attacking the neck of a stag (Pl. X).[39] At Clare Island the link between commemoration and hunting is made very clear. The stag hunt on the wall beside the tomb, coupled with the mounted warrior on the wall opposite, was surely calculated to enhance an appreciation of the lordly prowess of the deceased and his family, whoever they might have been.

While it is difficult to make much overall sense of the images on the ceiling, many of them seem to underline the theme of lordship. These include further paintings of horsemen, along with more scenes of hunting, the latter compositions including the vibrantly drawn greyhound. It is likely that the harpist inhabits the same milieu, for the music of the harp was an essential accompaniment for any self-respecting Gaelic lord. In fact one of the late medieval tomb slabs from Iona depicts a harpist, with a horseman alongside.[40] But what are we to make of the strange associations of man and beast that saunter across the ceiling at Clare Island, for example the

spear-bearer with an arrow in his back? It is certainly tempting to regard them as references to ancient battles and cattle raids, or illustrations perhaps of one of the great epics of the Irish past. Deciphering these and other paintings is not helped by the way in which the craftsmen virtually ignored the ribs as a potential framework, the scenes continuing from one cell to another.[41]

What is most disturbing to modern eyes is the way in which the sacred and profane are mixed together in an apparently arbitrary way. Hunting scenes were an obvious way of decorating the residences of Gaelic chiefs: the subject was depicted on the wall of the tower house at Urlan More (Co. Clare), and similar scenes are mentioned in a fifteenth-century poem that describes a house set within the bawn of Rudhraighe Mac Mathghamhna (Co. Monaghan) This was a timber building with painting 'on the surface of the brown oaken boards'. The poem describes a 'hidden hunt' with the likeness of stags and does, horsemen, hounds, decorative dogs and motionless bird flocks.[42] Such scenes were obviously popular with Gaelic chieftains, and it is likely that many an Irish tower house was embellished with colourful images of huntsmen, hounds and fleeing stags. The scheme described in the poem sounds remarkably like that at Clare Island, but in the latter case, amid the scenes of hunting, raiding and violence, space is found for some images with explicit Christian connotations, the pelican feeding her young, for example, one of the most popular metaphors of Christian salvation.[43] There is also the archangel Michael, holding the scales, an image conveying an unambiguous message about future judgement. In other cases the intention is not so clear-cut. Among the many depictions of animals and beasts—hare, wolf, heron, dragon, serpent, goats, pigs, etc.—is that of the cock, which might have found its way to the vault as one of the *arma Christi* or 'instruments' of the Passion. This would seem far-fetched, if it were not for the depiction of an isolated hand, another of the *arma Christi*, generally interpreted as the hand of one of Christ's tormentors (with or without whip).[44] From the fourteenth century the *arma Christi* played a fundamental role in medieval devotion, helping the believer to relive the suffering of Christ. It is difficult to understand why only two of the instruments were depicted at Clare Island and why they were not represented in a more coherent way. One gets the impression that the artists used them as isolated motifs, without appreciating their devotional significance. The *arma Christi* are frequently encountered in Irish art on funerary monuments, though usually in a

Pl. XI   Tomb slab from Ballysaggart, Co. Donegal (Royal Society of Antiquaries of Ireland).

limited cycle without the hand and the cock.[45] The heterogeneous collection of images at Clare Island, especially the interest in animals, suggests that the painters had at their disposal some sort of model sheet, a crude version perhaps of the sort of drawings found in the Pepysian model book of *c.* 1370–90, with its collections of beasts, animals and human figures.[46] Certainly the repertoire of Irish craftsmen seems to have been fairly limited, to judge from the repetition of subjects in tomb sculpture.

Other images at Clare Island that come from a 'standard' repertoire include the two-headed eagle and the somewhat emaciated pair of wrestlers. The two-headed eagle was a popular subject among the tile-makers, and examples are known from at least ten Irish sites, including the Cistercian monasteries of Mellifont and Dunbrody, where the subject occurs on line-impressed tiles, manufactured from the fourteenth century onwards.[47] Although there are parallels for the wrestlers in early Irish art, it was a common theme in the Gothic era. In addition to the well-known example in Villard de Honnecourt's sketchbook, the subject can be found on English misericords: there is a striking sixteenth-century example at Halsall (Lancashire).[48] The theme was familiar to Irish stone-carvers in the later middle ages, to judge from a sixteenth-century tomb slab from Ballysaggart (Co. Donegal). This is the so-called Mac Swiney slab, now at Killybegs (Pl. XI). The surface of the stone is divided into nine compartments, one with an armed warrior, another with two fighters locked in combat.[49]

The meaning of such images depends largely on their context. When wrestlers were carved on the Irish scripture crosses of the ninth and tenth centuries, they might have been intended to represent Jacob struggling with the angel,[50] but there is no reason to suppose that such readings endured throughout the later middle ages. It is far from obvious how an Irish audience would have interpreted such images in the fifteenth century. In the context of a tomb slab the wrestlers might have suggested physical prowess. One gets the impression that the theme belongs to a common stock of motifs exploited by local craftsmen with no particular intent, except perhaps to amuse and entertain. While compositions remained unchanged over centuries, meaning was far from immutable.

Although the paintings at Clare Island occasionally provide hints of the work of earlier centuries, there was little attempt at the sort of self-conscious revivalism seen on such works as Brian Boru's harp or the leather satchel of the Book of Armagh.[51] The wonderfully lithe drawing of the greyhound invites comparison with the springy cats in the Book of Kells, but for the most part the choice of subjects and the manner in which they are painted belong to the world of late Gothic Europe. The drawing of the hare and the hounds or even the dragons would not be out of place in the margins of a fifteenth-century book of hours, a place where one encounters that same ambiguous interplay of secular and religious iconography.[52] In fact lively animals can be found in Irish metalwork of the late Gothic era, on the Domhnach Airgid, for example, where the borders include a dog (?), the energetic outline of which recalls the hounds racing across the vault on Clare Island.[53] Given the decayed state of the paintings, it is difficult to assess their quality, but the small scale of the images on the vault and the lack of coherent composition gives the whole enterprise a decidedly rustic flavour. In this respect the work is a contrast to the bold and far more sophisticated paintings recently discovered at Ardamullivan (Co. Galway).[54] However, the individual images display considerable vitality, even if they are poorly integrated. By contrast, the paintings on the walls are larger and more impressive in scale: the stag fleeing between the pinnacles of the tomb was drawn with particular verve and confidence.

Remaining questions and enigmas

One conclusion that is hard to avoid is the fact that, with the exception of the simulated ribs, there is nothing to betray the Cistercian origin of the church. Even though the Cistercian connection somehow survived, it appears that by the end of the middle ages the chapel had become a monument to the chieftains of Clare Island rather than a spiritual outpost of the Cistercian order.

But who was the Gaelic lord (or lords) who transformed the church into what was virtually a private mausoleum? As Conleth Manning has pointed out, one seventeenth-century source states that the chapel was built by Diarmait Ó Máille (O'Malley), the founder of the friary at Murrisk, who died in the mid-1450s (Manning, this volume, p. 10). The style and subject matter of the paintings would not be inconsistent with such a date, but the presence of the ornate tomb provides reason for doubt. In view of its connections with the monument at Strade, the tracery, which must have preceded the painted decoration on the adjoining wall, appears to fit more comfortably into the context of the early sixteenth century. For the time being, therefore, the date of the late medieval embellishment of the church, along with the identity of the patron or patrons responsible, remains an open question.

Acknowledgements

In preparing this chapter I am grateful for the help and advice I have received from Katharine Simms, Conleth Manning, Karena Morton, Christoph Oldenbourg, Madeleine Katkov, Michael O'Neill, Peter Harbison and Rachel Moss.

NOTES

1. W. O'Sullivan (ed.), *The Strafford inquisition of County Mayo* (Dublin, Irish Manuscripts Commission, 1958), 39. The key passage dating from 1635, together with other sources, is cited in N. Ó Muraíle, 'The place-names of Clare Island', in C. Mac Cárthaigh and K. Whelan (eds), *New Survey of Clare Island. Volume I: history and cultural landscape* (Dublin, Royal Irish Academy, 1999), 104.

2. The extent of the land that went with the Abbey is discussed by C. Manning, 'Some notes on Clare Island in the 16th century', in *New Survey of Clare Island Newsletter* 5 (2002), 7–8.

3. Ware owned the ancient chartulary of Abbeyknockmoy (now lost), and this is presumably where he got his information about Clare Island.

4. The term is much used by A. Gwynn and R.N. Hadcock, *Medieval religious houses, Ireland* (London, Longman, 1970), *passim*, chiefly to describe failed houses.

5. E.A. D'Alton, *History of the Archdiocese of Tuam* (2 vols, Dublin, Phoenix, 1928), vol. 1, 209–10. Elsewhere D'Alton makes the assumption that the church on Clare Island was founded 'after consultation with Crovderg and for the purpose of appointing a place of penance and exile for members of the Cistercian Order' (D'Alton, *History of the Archdiocese of Tuam*, vol. 2, 337–9).

6. For place names see Ó Muraíle, 'Place-names of Clare Island', 99–141.

7. The archaeological surveys of the island have found no evidence for any other church site, in the absence of which one has to assume that the existing monument served the local populace. There are however hints of other churches in place names: Port na Cilleadh, the port of the church; An Chill Bhig, the small church (Ó Muraíle, 'Place-names of Clare Island', 122).

8. Abbeyknockmoy was founded by the Cathal Crobderg Ó Conchobair in 1190, and in its early years the abbey was closely identified with the O'Conor family, R. Stalley, *The Cistercian monasteries of Ireland* (London and New Haven, Yale University Press, 1987), 240.

9. John O'Donovan (ed.), *Annals of the kingdom of Ireland by the four masters*, (7 vols, Dublin, 1848–53), *sub anno* 1224.

10. The history of Mayo in this era is outlined by H.T. Knox, *The history of the county of Mayo* (Dublin, Hodges Figgis and Co., 1908), 62–89. The relationships between the O'Conors as overlords and the O'Malleys was at times fraught; in 1220, for example, it is reported that 'Dubhdara, son of Muiredhach O'Maille, was killed in a dispute by Cathal Crobhderg, in his own camp, in violation of all Connacht; and this was a grievous act, although it was his own misdeeds that recoiled on him', W.M. Hennessy, *The annals of Loch Cé* (London, Rolls Series, 1871), *sub anno* 1220. For an insight into the politics of Mayo at this time see K. Simms, 'A lost tribe—the Clan Murtagh O'Conors', *Journal of the Galway Archaeological Society* 53 (2001), 3–6.

11. Knox, *History of Mayo*, 237; J.F. Quinn, *History of Mayo* (Ballina, Brendan Quinn, 1993), 240, 243. There is no reference to either Kilgeever or Clare Island in the 1302–6 taxation records, but the vicarage of Kyllgayvayr is mentioned in Bishop Bodkin's Visitation (1558–9), H.T. Knox, *Notes on the early history of the dioceses of Tuam, Killala and Achonry* (Dublin, Hodges Figgis, 1904), 207.

12. Deliberate repetition of architectural features as a way of displaying ownership is not unknown in the middle ages. The twelfth-century church at Lindisfarne, for example, echoes many of the features of its mother house at Durham Cathedral; while this might be seen as a matter of contemporary fashion, it must also have reinforced the bond between the communities.

13. The location on the north side of the sanctuary remained a place of honour throughout the middle ages. Among many examples is the tomb of Rahere, the ancestral founder of the church of St Bartholomew, Smithfield, London, which dates to *c.* 1405. See B. Cherry, 'Some new types of late medieval tombs in the London area', in L. Grant (ed.), *Medieval art, architecture and archaeology in London, British Archaeological Association conference transactions for the year 1984* (London, British Archaeological Association Conference Transaction Series, 1990), 143.

14. Parallels outside Ireland include two reconstructed gable tombs in the Lady Chapel at St Davids Cathedral. One with traceried oculi now forms part of a memorial to Bishop John Owen (d. 1926). It is largely the work of Oldrid Scott, but based on medieval fragments. A second ornate tomb on the south wall, also reinstated by Oldrid Scott and W.D. Caroe, reuses medieval pieces. This example includes a tall gabled canopy with tracery motifs based on the pointed trilobe. As the apex of the monument coincides with the springing of the vault ribs, the original monument was evidently designed as part of the remodelling of a chapel *c.* 1500.

15. The group is discussed in H.G. Leask, *Irish churches and monastic buildings*, (3 vols, Dundalk, Dundalgan Press, 1955–60), vol. 3, 167–74. It is worth noting that the west doorway of the former friary at Ballyhaunis (Mayo) is surmounted with a screen of tracery (recently restored); this has affinities with some of the western tombs for it includes a segmental arch with hanging cusps, as well as a vertical 'mullion' rising above.

16. T. McNeill, 'The archaeology of Gaelic lordship east and west of the Foyle', in P.J. Duffy, D. Edwards and E. Fitzpatrick (eds), *Gaelic Ireland: land, lordship and settlement c. 1250–c. 1650* (Dublin, Four Courts Press, 2001), 348–51, has argued the case for the late fourteenth century. Although the tomb at Dungiven had previously been associated with Cooey na Gall O'Cahan (d. 1385) and ascribed to the end of the fourteenth century, the date was rejected by J. Hunt, *Irish medieval figure sculpture 1200–1600* (2 vols, Dublin and London, Irish University Press / Sotheby Parke Bernet, 1974), vol. 1, 131. Hunt's view gets some support from the design of the tracery with rotating mouchettes; this has affinities with Scottish work of the fifteenth century, in particular with the windows added to the abbey church at Iona, work ascribed to the middle of the century. See the Royal Commission on the Ancient and Historical Monuments of Scotland (hereafter cited as RCAHMS), *Argyll, an inventory of the monuments, vol. 4, Iona* (Edinburgh, RCAHMS, 1982), 52, 109. The name of the craftsman involved was Donaldus O'Brolchan, a family name with ancient links with Derry. A. Hamlin, 'Dungiven Priory and the Ó Catháin family' in M. Richter and J.-M. Picard

NOTES

(eds), *Ogma: essays in Celtic Studies in honour of Próinséas Ní Chatháin* (Dublin, Four Courts Press, 2002), 129, accepted a fifteenth century date for the tomb.

17. Leask, *Irish churches and monastic buildings*, vol. 3, 167. It is worth noting that a tomb chest at Athenry has similar arcading, for which Hunt, *Irish medieval figure sculpture*, vol. 1, 148, suggests an early sixteenth century date.

18. Hunt, *Irish medieval figure sculpture*, vol. 1, 150–1; R.A. Stalley, 'Maritime pilgrimage from Ireland and its artistic repercussions' in V. Almazán, B. Tate and M. Domínguez (eds), *Rutas Atlánticas de Peregrinacíon a Santiago de Compostela, Actas del il Congreso Internacional de Estudios Jacobeos* (Ferrol, Xunta de Galicia, 1999), 255–75.

19. The tomb at Strade was examined at length by L. Henry, 'A study of the canopied tomb niche in Strade Friary, County Mayo' (Unpublished BA dissertation, Department of the History of Art, University of Dublin, Trinity College, 1999).

20. The carvings appear to record visits to Canterbury (Becket), Cologne (Three Magi) and Rome (St Peter and St Paul).

21. The Ormond 'school' was examined by E. Rae, 'Irish sepulchral monuments of the later middle ages', *Journal of the Royal Society of Antiquaries of Ireland* 100 (1970). Henry, in 'Canopied tomb niche in Strade Friary', explored the sculptural connections with the Ormond school and came to the conclusion that the Strade monument belonged to the early or middle years of the sixteenth century.

22. Leask, *Irish churches and monastic buildings*, 169. The 1561 date is based on evidence in an account of the Lynch family written *c*. 1820. This states that the 'Lynch aisle' was erected by Nicholas Lynch fitz Stephen, and that it formed one of the transepts. See M.J. Blake, 'Account of the Lynch Family', *Journal of the Galway Archaeological and Historical Society* 8 (1913–14), 76–93, especially 90. Although the nineteenth century author had access to ancient records, there must be some doubt about the reliability of the evidence. Hunt, *Irish medieval figure sculpture*, vol. 1, 149, ignored the 1561 date and without explanation assigned the Joyce tomb to the late fifteenth century. It is of course possible that the monument was moved from another part of the church after the transept was complete. The whole issue warrants further investigation.

23. S. Simpson, 'The choir stalls and rood screen', in J. Geddes (ed.), *King's College Chapel Aberdeen, 1500–2000* (Leeds, Northern Universities Press, 2000), 74–97. The screen at Carlisle carries the monogram of Prior Gondibur (1484–1507); the parallels with the Gondibur screen at Carlisle were pointed out by Danielle O'Donovan on the occasion of the annual conference of the British Archaeological Association in July 2001.

24. See the many examples illustrated in D.H. Williams, *Catalogue of seals in the National Museum of Wales, Vol. I, Seal dies, Welsh seals and Papal bullae* (Cardiff, National Museum of Wales, 1993). There are also Irish versions of the conventional English equestrian portraits, as reflected in a seal matrix belonging to Mac Conmara of Clare (1312–28), R. ÓFloinn, 'Goldsmiths' work in

Ireland, 1200–1400', in C. Hourihane (ed.) *From Ireland coming: Irish art from the early Christian to the late Gothic period and its European context* (New Jersey, Princeton University Press, 2001), 301–3.

25. Edmund in fact is praying while on horseback, P. Binski, *Westminster Abbey and the Plantagenets* (New Haven and London, Yale University Press, 1995), 117; C. Wilson *et al.*, *Westminster Abbey* (London, New Bell's Cathedral Guides, 1986), 132–3.

26. Hunt, *Irish medieval figure sculpture*, vol. 1, 150. Hunt dated the Glinsk figure to the first quarter of the sixteenth century.

27. The most useful guide to Gaelic armour is P. Harbison, 'Native Irish arms and armour in medieval Gaelic literature, 1170–1600', *Irish Sword* 12 (1976) (48), 173–99, (49), 270–84.

28. Trinity College Dublin, MS 1440, fol. 24r. Given the way in which the outer coat is depicted, it could be argued that the Clare Island rider is wearing a quilted aketon, as worn by the lords of the Western Isles, K.A. Steer and J.W.M. Bannerman, *Late medieval monumental sculpture in the west Highlands* (Edinburgh, RCAHMS, 1977), 22–9. However the Irish evidence, both documentary and visual, makes it likely that some form of hauberk was intended.

29. Translation by Mahaffy and cited in Harbison, 'Native Irish arms', 176.

30. Harbison, 'Native Irish arms', fig. 1 facing p.180; Ó Floinn, 'Goldsmiths' work in Ireland', #295; R. Ó Floinn, 'Domnach Airgid shrine' in M. Ryan (ed.) *Treasures of Ireland* (Dublin, Royal Irish Academy, 1983), 176–7.

31. British Library MS Harley 1319. The illuminations are the work of a Parisian artist and were painted between 1401 and 1405, G. Mathew, *The court of Richard II* (London, John Murray, 1968), 209–10; E.M. Thompson, 'A contemporary account of the fall of Richard the Second', *Burlington Magazine* 5 (1904), 16–72, 267–78; M. Meiss, *French painting in the time of Jean de Berry: the Limbourgs and their contemporaries* (London, George Braziller, 1974).

32. J. Webb, 'Transactions of a French metrical history of the deposition of King Richard II', *Archaeologia*, 20 (1824), 161.

33. It is perhaps possible that Creton's original text included some form of accompanying drawings.

34. S. Mac Mathúna, 'An inaugural ode to Hugh O'Connor (king of Connacht 1293–1309)', *Zeitschrift für Celtische Philologie* 49–50 (1998), 563.

35. There appears to be no formal publication of the Annagh relief: it could be dated at any period between the thirteenth and the sixteenth century. It is possible that the figure is holding a sickle rather than a spear, and as such was intended to be one of the four horsemen of the Apocalypse.

36. Steer and Bannerman, *Late medieval monumental sculpture*, pl. 6B.

37. J. Cummins, *The hound and the hawk* (London, Weidenfeld and Nicolson, 2001).

38. Hunting scenes are found on misericords at Beverley, Ely, Boston, Manchester, Gloucester, Bristol, Chester etc., J.D.C. Smith, *A guide to church woodcarvings, misericords*

NOTES

and bench-ends (Newton Abbot, David & Charles, 1974), 46; C. Grossinger, *The world upside down: English misericords* (London, Harvey Miller Publishers, 1997), 166–7. There is also an excellent example at New College Oxford, F.W. Steer, *Misericords at New College, Oxford* (Chichester, Phillimore, 1973), 14–15. Hunting is also depicted on floor tiles from the Cistercian abbey at Neath.

39. D.H. Caldwell, *Angels, noble and unicorns: art and patronage in medieval Scotland* (Edinburgh, National Museum of the Antiquities of Scotland, 1982), 54–5; Steer and Bannerman, *Late medieval monumental sculpture*, pl. 26C.

40. Steer and Bannerman, *Late medieval monumental sculpture*, pl. 5D.

41. In some instances the painting of the ribs appears to overlap the individual images, as if the ribs were an afterthought.

42. K. Simms, 'Native sources for Gaelic settlement: the house poems', in P.J. Duffy, D. Edwards and E. Fitzpatrick (eds), *Gaelic Ireland: land, lordship and settlement c. 1250–c. 1650* (Dublin, Four Courts Press, 2001), 252–4.

43. For a full discussion of the iconography of the pelican and its appearance in Irish art see C. Hourihane, *Gothic art in Ireland 1169–1550* (London and New Haven, Yale University Press, 2003), 99–113.

44. H. Van Os, *The art of devotion in the Late Middle Ages in Europe 1300–1500* (London and Amsterdam, Merrell Holberton, 1994), 114. Full sets of the *arma Christi* are found in depictions of the Mass of St Gregory. The isolated hand appears with the *arma Christi* on a bench end at Fressingfield (Suffolk); the fact that it appears alongside the jug (of Pilate) has led to suggestions that its represents the washing of hands, Smith, *A guide to church woodcarvings*, 29. While the hand may have been copied as one of the *arma Christi*, it is worth noting the uncertainty about the interpretation of this image (see Section Two). Furthermore, a hand forms part of the heraldry of the de Burgo family, as illustrated in TCD MS

1440, fol. 24r.

45. R. Roe, 'Instruments of the Passion: notes towards a survey of their illustration and distribution in Ireland', *Old Kilkenny Review* 2–5 (1983), 527–34. See also Hourihane, *Gothic art in Ireland*, 95, 134.

46. Cambridge, Magdalene College MS Pepys 1916; J.J.G. Alexander and P. Binski (eds), *Age of chivalry: art in Plantagenet England 1200–1400*, in Exhibition catalogue (Royal Academy of Arts, London, 1987), 402; Grossinger, *The world upside down*, 58.

47. E. Eames and T. Fanning, *Irish medieval tiles* (Dublin, Royal Irish Academy, 1988), 87, 119.

48. Grossinger, *The world upside down*, 172; Smith, *A guide to church woodcarvings*, 46.

49. Hunt, *Irish medieval figure sculpture*, vol. 1, 133.

50. P. Harbison, *The high crosses of Ireland*, (3 vols, Bonn, Habelt on behalf of the Römisch-Germanisches Zentralmuseum and the Royal Irish Academy, 1992), vol. 1, 237–8, discusses this image and raises doubts about its identification with Jacob and the angel.

51. Both objects are housed in the Library of Trinity College Dublin; see P. Cone, *Treasures of early Irish art 1500 BC to 1500 AD* (New York, Metropolitan Museum of Art, 1978), 217–9, pls 67–8. For a recent discussion of the Gaelic revival in the later middle ages see Hourihane, *Gothic art in Ireland*, 139–52.

52. There is a wonderful parallel for the intermixing of secular and Christian imagery in the tomb of Alexander MacLeod (d. 1547) at Rodel on the Isle of Harris (made in 1528), which shows carvings of huntsmen with three dogs, a group of stags, a castle and a ship, together with images of St Michael and the weighing scales, plus two bishops and the Virgin Mary, Steer and Bannerman, *Late medieval monumental sculpture*, 97–8, Pls 31–2.

53. Ó Floinn, 'Domnach Airgid shrine', 176–7; Ó Floinn, 'Goldsmiths' work in Ireland', 71.

54. K. Morton, 'Medieval wall paintings at Ardamullivan', *Irish Arts Review Yearbook* 18 (2002), 104–13.

BIBLIOGRAPHY

Alemand, L.A. 1690 *Histoire monastique d'Irlande*. Paris. Michel Guerout.

Alemand, L.A. 1722 *Monasticon Hibernicum, or the monastical history of Ireland*. J. Stevens (trans.). London. William Mears.

Alexander, J.J.G. and Binski, P. (eds) 1987 *Age of chivalry: art in Plantagenet England 1200–1400*. Exhibition catalogue. Royal Academy of Arts. London.

Anderson, M.D. 1971 *History and imagery in British churches*. Edinburgh. John Murray.

Ashe Fitzgerald, M. 2000 *Thomas Johnson Westropp (1860–1922): an Irish antiquary*. Seandálaíocht: Mon. 1. Dublin. Department of Archaeology, University College, Dublin.

Berry, H.F. 1898 *Register of wills and inventories of the diocese of Dublin in the time of archbishops Tregury and Walton 1457–1483*. Dublin. Dublin University Press for the Royal Society of Antiquaries of Ireland.

Bicknell, S. 1996 *The history of the English organ*. Cambridge. Cambridge University Press.

Binski, P. 1995 *Westminster Abbey and the Plantagenets*. New Haven and London. Yale University Press.

Blake, M.J. 1913–14 Account of the Lynch Family. *Journal of the Galway Archaeological and Historical Society* **8**, 76–93.

Bläuer Böhm, C. 1994 Mineralogical examination. Unpublished report, Fachhochschule Köln, BMFT-Projekt.

Bowles, E.A. 1982 A preliminary checklist of fifteenth-century organs in paintings and manuscript illuminations. *The Organ Yearbook* **13**, 5–30.

Boydell B. 1986 Music before 1700. In T.W. Moody and W.E. Vaughan (eds), *A new history of Ireland*, IV, 542–628. Oxford. Clarendon Press.

Boydell, B. (ed.) 1999 *Music at Christ Church before 1800: documents and selected anthems*. Dublin. Four Courts Press.

Bromwich, R. 1961 Celtic dynastic themes and the Breton lays. *Études Celtiques* **9**, 439–74.

Buckley, A. 1978 What was the tiompán? A problem in ethnohistorical organology: evidence in Irish literature. *Jahrbuch für musikalische Volks- und Völkerkunde* **9**, 53–88.

Buckley, A. 1990 Musical instruments in Ireland from the 9th to the 14th centuries: a review of the organological evidence. In G. Gillen and H. White (eds), *Irish musical studies 1*, 13–57. Blackrock. Irish Academic Press.

Buckley, A. 1995 Music-related imagery on early Christian insular sculpture: identification, context, function. *Imago Musicae/International Yearbook of Musical Iconography* **8** (1991), 135–99.

Buckley, A. 1997 Music and manners: readings of medieval Irish literature. *Bullán: an Irish Studies Journal* **3**(1), 33–43.

Buckley, A. 2001 Timpan/Tiompán in S. Sadie (ed.), *The New Grove dictionary of music and musicians*, vol. 25, 482–3. London. Macmillan.

Buckley, A. 2003 Representations of musicians in medieval Christian iconography of Ireland and Scotland as local cultural expression. In Katherine A. McIver (ed.), *Art and music in the early modern period. Essays in honor of Franca Trinchieri Camiz*, 217–31. Aldershot. Ashgate.

Burke, J. and Burke J.B. 1844 *Encyclopaedia of heraldry, or general armory of England, Scotland, and Ireland*. 3rd edn. London. Henry G. Bohn.

Caldwell, D.H. 1982 *Angels, noble and unicorns: art and patronage in medieval Scotland*. Edinburgh. National Museum of the Antiquities of Scotland.

Cantwell, I. 2002 *Memorials of the dead: counties Galway & Mayo (western seaboard)*. Irish Memorial Inscriptions, Vol. 1.

CD-ROM. Dublin. Eneclann Ltd.

Cherry, B. 1990 Some new types of late medieval tombs in the London area. In L. Grant (ed.), *Medieval art, architecture and archaeology in London. British Archaeological Association conference transactions for the year 1984*, 140–54. London. British Archaeological Association Conference Transaction Series.

Cochrane, R. 1901 Excursion to County Galway. *Journal of the Royal Society of Antiquaries of Ireland* **31**, 305–40.

Cone, P. 1978 *Treasures of early Irish art 1500 BC to 1500 AD*. New York. Metropolitan Museum of Art.

Corlett, C. 2001 *Antiquities of West Mayo*. Bray. Wordwell.

Crawford, H.S. 1915 Mural paintings at Holycross Abbey. *Journal of the Royal Society of Antiquaries of Ireland* **45**, 149–51.

Crawford, H.S. 1919 Mural paintings and inscriptions at Knockmoy Abbey, *Journal of the Royal Society of Antiquaries of Ireland* **49**, 25–34.

Cullen, C. and Gill, P. 1992 *Holy wells and Christian settlement on Clare Island, Co. Mayo*. Clare Island Series 4. Clare Island. Centre for Island Studies.

Cummins, J. 2001 *The hound and the hawk*. London. Weidenfeld and Nicolson.

D'Alton, E.A. 1928 *History of the Archdiocese of Tuam*, 2 vols. Dublin. Phoenix.

Deane, T.N. 1880 Clare Island. In appendix to *The 48th report of the Commissioners of Public Works in Ireland*, 75. HMSO.

Duffy, E. 1992 *The stripping of the altars: traditional religion in England c. 1400–c. 1580*. New Haven and London. Yale University Press.

Dunleavy, M. 1989 *Dress in Ireland*. London. Batsford Press.

Eames, E. and Fanning, T. 1988 *Irish medieval tiles*. Dublin. Royal Irish Academy.

Fitzgerald, W. 1895–7 Abbeyknockmoy, Co. Galway. *Journal of the Association for the Preservation of the Memorials of the Dead in Ireland* **3**, 277–8.

Fleischmann, A. 1952 References to chant in early Irish MSS. In S. Pender (ed.), *Féilscríbhinn Torna*, 43–9. Cork. Cork University Press.

Flood, W.H. Grattan 1910 Irish organ-builders from the eighth to the close of the eighteenth century. *Journal of the Royal Society of Antiquaries of Ireland* **40**, 229–34.

Flood, W.H. Grattan 1970 *A history of Irish music*, 3rd edn. Shannon. Irish University Press.

Freeman, A.M. (ed.) 1970 *Annála Connacht: the Annals of Connacht (A.D. 1224–1544)*. Dublin. The Dublin Institute for Advanced Studies.

Gillespie, F. 2000 Heraldry in Ireland: an introduction. *The Double Tressure: Journal of the Heraldic Society of Scotland* **23**, 7–25.

Gleeson, D.F. 1936 Drawing of a hunting scene, Urlan Castle, Co. Clare. *Journal of the Royal Society of Antiquaries of Ireland* **66**, 193.

Gosling, P., Manning, C. and Waddell J. (eds) forthcoming *New Survey of Clare Island. Volume 5: archaeology*. Dublin. Royal Irish Academy.

Gosling. P. 1993 *Archaeological inventory of County Galway. Volume 1: west Galway*. Dublin. Stationery Office.

Grabar, A. 1968 *Christian iconography: a study of its origins*. New Jersey. Princeton University Press.

Graham, J.R. 2001 The geology of Clare Island: perspectives and problems. In J.R. Graham (ed.), *New survey of Clare Island. Volume 2: geology*. Dublin. Royal Irish Academy.

Grindle, W.H. 1989 *Irish cathedral music: a history of music at the cathedrals of the Church of Ireland*. Belfast. Institute of Irish Studies, Queen's University Belfast.

Grossinger, C. 1997 *The world upside down: English misericords*. London. Harvey Miller Publishers.

Gwynn, A. and Hadcock, R.N. 1970 *Medieval religious houses, Ireland*. London. Longman.

Hamlin, A. 2002 Dungiven Priory and the Ó Catháin family. In M. Richter and J.-M. Picard (eds), *Ogma: essays in Celtic Studies in honour of Próinséas Ní Chatháin*, 118–37. Dublin. Four Courts Press.

Harbison, P. 1976 Native Irish arms and armour in medieval Gaelic literature, 1170–1600. *Irish Sword* **12** (48),173–99, (49) 270–84.

Harbison, P. 1992 *The high crosses of Ireland*, 3 vols. Bonn. Habelt, on behalf of the Römisch-Germanisches Zentralmuseum and the Royal Irish Academy.

Harbison, P. 1995 Architectural sculpture from the twelfth century at Ardmore. *Irish Arts Review* **11**, 96–102.

Hardiman, J. 1846 *A chorographical description of West or h-Iar Connaught written in A.D. 1684 by Roderick OFlaherty, Esq*. Dublin. The Irish Archaeological Society.

Hayden, J.A. 1969 *Misericords in St Mary's Cathedral, Limerick*. Revised by Rev. M.J. Talbot, Dean of Limerick, 1969. Limerick Leader.

Hayes-McCoy, G.A. 1996 *Scots mercenary forces in Ireland (1565–1603)*. 2nd edn. Dublin. Edmund Burke Publisher.

Hedges, R.E.M., Pettitt, P.B., Bronk Ramsey, C. and van Klinken, G.J. 1998 Radiocarbon dates from the Oxford AMS system archaeometry datelist 25, *Archaeometry* **40**, 227–39.

Hennessy, W.M. 1871 *The annals of Loch Cé*. London. Rolls Series.

Henry, F. 1960–61 Remarks on the decoration of three Irish psalters. *Proceedings of the Royal Irish Academy* **61**C, 23–40.

Henry, L., 1999 A study of the canopied tomb niche in Strade Friary, County Mayo. Unpublished BA dissertation. Department of the History of Art, University of Dublin, Trinity College.

Higgins, J. 1987 *The early Christian cross slabs, pillar stones and related monuments of county Galway, Ireland*, 2 vols. BAR International Series 375. London. HMSO.

Holmes, J. 1984 The organ in Ireland. Unpublished pamphlet.

Hourihane, C. 1984 *The iconography of religious art in Ireland, 1250–1550*. Unpublished PhD thesis at the Courtauld Institute of Art, University of London.

Hourihane, C. 2003 *Gothic art in Ireland 1169–1550*. London and New Haven. Yale University Press.

Hulme, E. 1975 *Symbolism in Christian art*. 2nd edn. Guildford. Biddles.

Hunt, J. 1974 *Irish medieval figure sculpture 1200–1600*, 2 vols. Dublin and London. Irish University Press/Sotheby Parke Bernet.

James, M.R. 1951 *Pictor in carmine*. *Archaeologia* **94**, 141–66.

Jongelinus, G. 1640 *Notitiae abbatiarum ordinis Cistertiencis per universum orbem*. Coloniae Agrippinae. Apud Ioannem Henningium Bibliopolam.

King, J.E. 1930 *Baedae Opera Historica*, vol. II, 405–7. London and New York. Harvard University Press.

Kirkpatrick, B. (ed.) 1992 *Brewers concise dictionary of phase and fable*. Helicon Publishing Ltd.

Knox, H.T. 1904 *Notes on the early history of the dioceses of Tuam, Killala and Achonry*. Dublin. Hodges Figgis and Co.

Knox, H.T. 1908 *The history of the county of Mayo*. Dublin. Hodges, Figgis and Co. Ltd.

Lacy, B. 1983 *Archaeological survey of County Donegal*. Lifford. Donegal County Council.

Leask, H.G. 1960 *Irish churches and monastic buildings. Volume 2: Gothic architecture to A.D. 1400*. Dundalk. Dundalgan Press.

Leask, H.G. 1960 *Irish churches and monastic buildings. Volume 3: medieval Gothic, the last phases*. Dundalk. Dundalgan Press.

Leask, H.G. 1977 *Irish castles*. Dundalk. Dundalgan Press.

Leask, H.G. and Macalister, R.A.S. 1946 Liathmore-Mochoemóg (Leigh), County Tipperary. *Proceedings of the Royal Irish Academy* **51**C, 1–14.

Leigh Fry, S. 1999 *Burial in medieval Ireland 900–1500*. Dublin. Four Courts Press.

Lewis, S. 1837 *A topographical dictionary of Ireland*, 2 vols. London. S. Lewis & Co.

Lucas, A.T. 1989 *Cattle in ancient Ireland*. Kilkenny. Boethius Press.

Luckombe, P. 1780 *A tour through Ireland*. J. Place and R. Byrn for Mssrs Whitestone, Sleater, Sheppard.

Mac Mathúna, S. 1998 An inaugural ode to Hugh O'Connor (king of Connacht 1293–1309). *Zeitschrift für Celtische Philologie* **49–50**, 548–75.

MacCaffrey, J. (ed.) 1907 *The Black Book of Limerick*. Dublin. M.H. Gill & Son.

Manning, C. 1998 Clonmacnoise Cathedral. In H. King (ed.), *Clonmacnoise studies vol. I: seminar papers 1994*. Dublin. Dúchas.

Manning, C. 2002 Some notes on Clare Island in the 16th century. *New Survey of Clare Island Newsletter* **5**, 7–8.

Mathew, G. 1968 *The court of Richard II*. London. John Murray.

McKinnon, J. 1974 The tenth-century organ at Winchester. *The Organ Yearbook* **5**, 4–19.

McNeill, T. 2001 The archaeology of Gaelic lordship east and west of the Foyle. In P.J. Duffy, D. Edwards and E. Fitzpatrick (eds), *Gaelic Ireland: land, lordship and settlement c. 1250–c. 1650*, 346–56. Dublin. Four Courts Press.

Meiss, M. 1974 *French painting in the time of Jean de Berry: the Limbourgs and their contemporaries*. London. George Braziller.

Morton, K. 2001 Medieval wall paintings at Ardamullivan. *Irish Arts Review Yearbook* **18**, 104–13.

Morton, K. 2004 Irish medieval wall painting. *Medieval Lectures, the Barryscourt Lectures*, 311–49. Kinsale. Gandon Editions.

Morton, K. forthcoming Wall painting report. In Miriam Clyne (ed.), *Excavations at Kells Priory, Co. Kilkenny*. Dublin. Stationery Office.

Mulloy, S. 1988 *O'Malley people and places*. Whitegate and Westport. Ballinakella Press and Carrowbawn Press.

Murphy, D. (ed.) 1891 *Triumphalia chronologica Monasterii Sanctae Crucis in Hibernia*. Dublin. Sealy, Bryers & Walker.

Neary, D.M. 1997 Organ-building in seventeenth- and eighteenth-century Dublin, and its English connections. *The British Institute of Organ Studies Journal* **32**, 20–7.

Nicholls, K.W. 1983 Late medieval Irish annals: two fragments. *Peritia* **2**, 87–102.

O'Donovan, J. (ed.) 1854 *Annals of the kingdom of Ireland by the four masters*, 7 vols. Dublin.

O'Donovan, J. (ed.) 1856 *The Annals of the Kingdom of Ireland by the Four Masters* [Annála Ríoghachta Éireann]. 7 vols. Dublin. Hodder, Smith and Co.

O'Flanagan, M. (ed.) 1927 Ordnance survey name books Co. Mayo no. 100: from Kilcommon to Kilgeever. Unpublished typescript in Hardiman Library, National University of Ireland, MS D15.54.

Ó Floinn, R. 1983 Domnach Airgid Shrine. In M. Ryan (ed.) *Treasures of Ireland*, 176–7. Dublin. Royal Irish Academy.

Ó Floinn, R. 2001 Goldsmiths' work in Ireland, 1200–1400. In C. Hourihane (ed.) *From Ireland coming: Irish art from the early Christian to the late Gothic period and its European context*, 289–312. New Jersey. Princeton University Press.

O'Grady, S.H. 1926 *Catalogue of Irish manuscripts in the British Museum*, vol. 1. London. Printed for the Trustees of the British Museum.

Ó hÉailidhe, P. 1987 The cross-base at Oldcourt, near Bray, Co. Wicklow. In E. Rynne (ed.), *Figures from the past: studies on figurative art in Christian Ireland, in honour of Helen M. Roe*, 98–110. Dun Laoghaire. Grendale Press and Royal

Society of Antiquaries of Ireland.

Ó Muraíle, N. 1998 A description of County Mayo *c.* 1684 by R. Downing. In T. Barnard, D. Ó Cróinín and K. Simms (eds), *'A miracle of learning': studies in manuscripts and Irish learning. Essays in honour of William O'Sullivan*, 236–65. Aldershot. Ashgate.

Ó Muraíle, N. 1999 The place-names of Clare Island. In C. Mac Cárthaigh and K. Whelan (eds), *New Survey of Clare Island. Volume 1: history and cultural landscape*, 99–141. Dublin. Royal Irish Academy.

O'Sullivan, W. (ed.) 1958 *The Strafford inquisition of County Mayo.* Dublin. Stationery Office.

O'Sullivan, W. 1997 A finding list of Sir James Ware's manuscripts. *Proceedings of the Royal Irish Academy* **97**C, 69–99.

Otway, C. 1839 *A tour of Connaught: comprising sketches of Clonmacnoise, Joyce Country, and Achill.* Dublin. William Curry, Jun. and Company.

Park, D. 1986 Cistercian wall painting and panel painting. In D. Park and C. Norton (eds), *Cistercian art and architecture in the British Isles*, 181–210. Cambridge. Cambridge University Press.

Petersen, K. 1992 Microbiological examination, description of species, effects of UV-C irradiation. Unpublished report, Universität Oldenburg.

Provincia Autonoma di Trento 2002 *Catalogo a cura di Enrico Castelnuovo, Francesca de Gramatica: Il Gotico nelle Alpi 1350–1450.*

Quinn, J.F. 1993 *History of Mayo.* Ballina. Brendan Quinn.

Rae, E. 1970 Irish sepulchral monuments of the later middle ages. *Journal of the Royal Society of Antiquaries of Ireland* **100**, 1–38.

Rae, E.C. 1966 The sculpture of the cloister of Jerpoint Abbey. *Journal of the Royal Society of Antiquaries of Ireland* **96**, 59–91

Raschle, P. 1996 Microbiological examination, sequence of colonisation, preventative measures. Unpublished report, EMPA Eidgenössische MaterialPrüfungsAnstalt.

RCAHMS 1982 *Argyll, an inventory of the monuments, vol. 4, Iona.* Edinburgh. Royal Commission on the Ancient and Historical Monuments of Scotland.

Rietstap, J.-B. 1967 *Illustrations to the Armorial Général by V. and H.V. Rolland.* London. Heraldry Today.

Rimmer, J. 1984 *The Irish harp.* Cork. Mercier Press.

Roe, H.M. 1968 *Medieval fonts of Meath.* Meath Archaeological and Historical Society.

Roe, H.M. 1975 The cult of St. Michael in Ireland. In C. Ó Danachair (ed.), *Folk and farm: essays in honour of A.T. Lucas*, 251–64. Dublin. Royal Society of Antiquaries.

Roe, R. 1983 Instruments of the Passion: notes towards a survey of their illustration and distribution in Ireland. *Old Kilkenny Review* **2–5**, 527–34.

Rouse, C. 1972 *Medieval wall painting.* Risborough. Shire Publications.

Sekules, V. 2001 *Medieval art.* Oxford University Press.

Simington, R.C. (ed.) 1956 *Books of survey and distribution. Volume 2: county of Mayo.* Dublin. Stationery Office.

Simms, K. 2001 A lost tribe—the Clan Murtagh O'Conors. *Journal of the Galway Archaeological Society* **53**, 1–22.

Simms, K. 2001 Native sources for Gaelic settlement: the house poems. In P.J. Duffy, D. Edwards and E. Fitzpatrick (eds), *Gaelic Ireland: land, lordship and settlement c. 1250–c. 1650*, 246–67. Dublin. Four Courts Press.

Simpson, S. 2000 The choir stalls and rood screen. In J. Geddes (ed.), *King's College Chapel Aberdeen, 1500–2000*, 74–97. Leeds. Northern Universities Press.

Smith, J.D.C. 1974 *A guide to church woodcarvings, misericords and bench-ends.* Newton Abbot. David & Charles.

Stalley, R. 1987 *The Cistercian monasteries of Ireland.* London and New Haven. Yale University Press.

Stalley, R.A. 1999 Maritime pilgrimage from Ireland and its

artistic repercussions. In V. Almazán, B. Tate and M. Domínguez (eds), *Rutas Atlánticas de Peregrinacíon a Santiago de Compostela, Actas del il Congreso Internacional de Estudios Jacobeos*, 255–75. Ferrol. Xunta de Galicia.

Steer, F.W. 1973 *Misericords at New College, Oxford.* Chichester. Phillimore.

Steer, K.A. and Bannerman, J.W.M. 1977 *Late medieval monumental sculpture in the west Highlands.* Edinburgh. Royal Commission on Ancient and Historical Monuments of Scotland.

Stokes, W. (ed.) 1897 Cóir Anmann. *Irische Texte* **3**, 2 Heft, 317–23).

Stokes, W. (ed.) 1903 Echtra Mac Echach Muigmedóin. *Revue Celtique* **23**, 190–207.

Swan, L. 1983 Enclosed ecclesiastical sites and their relevance to settlement patterns of the first millennium A.D. In T. Reeves-Smyth and F. Hammond (eds), *Landscape archaeology in Ireland*, 269–94. British Series 116. Oxford. Archaeopress.

Thompson, E.M. 1904 A contemporary account of the fall of Richard the Second. *Burlington Magazine* **5**, 16–72, 267–78.

Van Os, H. 1994 *The art of devotion in the Late Middle Ages in Europe 1300–1500.* London and Amsterdam. Merrell Holberton.

Vigors, P.D. 1890 Reports from counties: County Mayo. *Journal of the Association for the Preservation of the Memorials of the Dead, Ireland* **1**, no. 1, 24.

Vigors, P.D. 1893 Reports from counties: County Mayo. *Journal of the Association for the Preservation of the Memorials of the Dead, Ireland* **1**, no. 4, 452–7.

Waddell, J. 1994 The archaeology of Aran. In J. Waddell, J.W. O'Connell and A. Korff (eds), *The book of Aran, the Aran Islands, Co. Galway*, 75–135. Kinvara. Tir Eolas.

Wallace, J.N.A. 1936–41 Frescoes at Urlanmore Castle, Co. Clare. *North Munster Antiquaries Journal* **1–2**, 38–9.

Ware, J. 1626 *Archiepiscoporum Casseliensium et Tuamensium vitae; duobus expressae commentariolis. Quibus adjicitur historia coenobiorum Cisterciensium Hiberniae.* Dublin. Ex Officina Societatis Bibliopolarum.

Ware, J. 1658 *De Hibernia et antiquitatibus eius disquisitiones.* 2nd edn. London. Jo. Crook.

Watters, P. 1872–3 Original documents connected with Kilkenny. *Journal of the Royal Society of Antiquaries of Ireland* **12**, 532–43.

Webb, J. 1824 Transactions of a French metrical history of the deposition of King Richard II. *Archaeologia* **20**, 1–423, pl. 4.

Went, A.E.J. 1952 Irish fishing spears. *Journal of the Royal Society of Antiquaries of Ireland* **82**, 109–34.

Westropp, T.J. 1911 Clare Island Survey: history and archaeology. *Proceedings of the Royal Irish Academy* **31** (1911–15), section 1, part 2, 1–78.

Whelan, K. 1999 Landscape and society on Clare Island 1700–1900. In C. Mac Cárthaigh and K. Whelan (eds), *New Survey of Clare Island. Volume 1: history and cultural landscape*, 73–98. Dublin. Royal Irish Academy.

White, N.B. (ed.) 1943 *Extents of Irish monastic possessions, 1540–1541, from manuscripts in the Public Record Office, London.* Coimisiún Láimhscríbhinní na hÉireann / Irish Manuscripts Commission. Dublin. The Stationery Office.

Williams, D.H. 1993 *Catalogue of seals in the National Museum of Wales, Vol. I, Seal dies, Welsh seals and Papal bullae.* Cardiff. National Museum of Wales.

Williams, P. and Owen, B. 1988 *The New Grove organ.* London and Basingstoke. Macmillan.

Wilson, C., Gem, R., Tudor-Craig, P., Physick, J. 1986 *Westminster Abbey.* London. New Bell's Cathedral Guides.

INDEX